SPECIFIC CONSIDERATIONS AND GUIDANCE FOR THE ESTABLISHMENT OF IONIZING RADIATION FACILITIES

The following States are Members of the International Atomic Energy Agency:

AFGHANISTAN	GERMANY	PALAU
ALBANIA	GHANA	PANAMA
ALGERIA	GREECE	PAPUA NEW GUINEA
ANGOLA	GRENADA	PARAGUAY
ANTIGUA AND BARBUDA	GUATEMALA	PERU
ARGENTINA	GUYANA	PHILIPPINES
ARMENIA	HAITI	POLAND
AUSTRALIA	HOLY SEE	PORTUGAL
AUSTRIA	HONDURAS	QATAR
AZERBAIJAN	HUNGARY	REPUBLIC OF MOLDOVA
BAHAMAS	ICELAND	ROMANIA
BAHRAIN	INDIA	RUSSIAN FEDERATION
BANGLADESH	INDONESIA	RWANDA
BARBADOS	IRAN, ISLAMIC REPUBLIC OF	SAINT KITTS AND NEVIS
BELARUS	IRAQ	SAINT LUCIA
BELGIUM	IRELAND	SAINT VINCENT AND
BELIZE	ISRAEL	THE GRENADINES
BENIN	ITALY	SAMOA
BOLIVIA, PLURINATIONAL	JAMAICA	SAN MARINO
STATE OF	JAPAN	SAUDI ARABIA
BOSNIA AND HERZEGOVINA	JORDAN	SENEGAL
BOTSWANA	KAZAKHSTAN	SERBIA
BRAZIL	KENYA	SEYCHELLES
BRUNEI DARUSSALAM	KOREA, REPUBLIC OF	SIERRA LEONE
BULGARIA	KUWAIT	SINGAPORE
BURKINA FASO	KYRGYZSTAN	SLOVAKIA
BURUNDI	LAO PEOPLE'S DEMOCRATIC	SLOVENIA
CAMBODIA	REPUBLIC	SOUTH AFRICA
CAMEROON	LATVIA	SPAIN
CANADA	LEBANON	SRI LANKA
CENTRAL AFRICAN	LESOTHO	SUDAN
REPUBLIC	LIBERIA	SWEDEN
CHAD	LIBYA	SWITZERLAND
CHILE	LIECHTENSTEIN	SYRIAN ARAB REPUBLIC
CHINA	LITHUANIA	TAJIKISTAN
COLOMBIA	LUXEMBOURG	THAILAND
COMOROS	MADAGASCAR	TOGO
CONGO	MALAWI	TONGA
COSTA RICA	MALAYSIA	TRINIDAD AND TOBAGO
CÔTE D'IVOIRE	MALI	TUNISIA
CROATIA	MALTA	TÜRKİYE
CUBA	MARSHALL ISLANDS	TURKMENISTAN
CYPRUS	MAURITANIA	UGANDA
CZECH REPUBLIC	MAURITIUS	UKRAINE
DEMOCRATIC REPUBLIC	MEXICO	UNITED ARAB EMIRATES
OF THE CONGO	MONACO	UNITED KINGDOM OF
DENMARK	MONGOLIA	GREAT BRITAIN AND
DJIBOUTI	MONTENEGRO	NORTHERN IRELAND
DOMINICA	MOROCCO	UNITED REPUBLIC
DOMINICAN REPUBLIC	MOZAMBIQUE	OF TANZANIA
ECUADOR	MYANMAR	UNITED STATES OF AMERICA
EGYPT	NAMIBIA	URUGUAY
EL SALVADOR	NEPAL	UZBEKISTAN
ERITREA	NETHERLANDS	VANUATU
ESTONIA	NEW ZEALAND	VENEZUELA, BOLIVARIAN
ESWATINI	NICARAGUA	REPUBLIC OF
ETHIOPIA	NIGER	VIET NAM
FIJI	NIGERIA	YEMEN
FINLAND	NORTH MACEDONIA	ZAMBIA
FRANCE	NORWAY	ZIMBABWE
GABON	OMAN	
GEORGIA	PAKISTAN	

The Agency's Statute was approved on 23 October 1956 by the Conference on the Statute of the IAEA held at United Nations Headquarters, New York; it entered into force on 29 July 1957. The Headquarters of the Agency are situated in Vienna. Its principal objective is "to accelerate and enlarge the contribution of atomic energy to peace, health and prosperity throughout the world".

RADIATION TECHNOLOGY SERIES No. 7

SPECIFIC CONSIDERATIONS AND GUIDANCE FOR THE ESTABLISHMENT OF IONIZING RADIATION FACILITIES

INTERNATIONAL ATOMIC ENERGY AGENCY
VIENNA, 2023

COPYRIGHT NOTICE

© IAEA, 2023

Printed by the IAEA in Austria
April 2023
STI/PUB/1999

IAEA Library Cataloguing in Publication Data

Names: International Atomic Energy Agency.
Title: Specific considerations and guidance for the establishment of ionizing radiation facilities / International Atomic Energy Agency.
Description: Vienna : International Atomic Energy Agency, 2023. | Series: IAEA radiation technology series, ISSN 2220–7341 ; no. 7 | Includes bibliographical references.
Identifiers: IAEAL 23-01576 | ISBN 978–92–0–147222–9 (paperback : alk. paper) | ISBN 978–92–0–147322–6 (pdf) | ISBN 978–92–0–147422–3 (epub)
Subjects: LCSH: Ionizing radiation — Safety measures. | Radiation — Safety measures. | Nuclear facilities.
Classification: UDC 613.648.4 | STI/PUB/1999

FOREWORD

Various nuclear techniques and irradiation services provide beneficial opportunities to enhance the quality of life by making use of ionizing radiation facilities. National nuclear institutions (NNIs), universities, medical centres and private companies have established and used IRFs not only for research and development (R&D) purposes but also for the provision of commercial services and goods. As the number of deployments of IRFs continues to increase, the importance of professional management of ionizing radiation facility projects has to be highlighted, in particular for those institutions and organizations with little or no experience in the area. Indeed, the ionizing radiation facility project needs to be planned, managed and conducted in such a way that it guarantees successful progress of its implementation and full utilization after the facility begins to operate and provide services. In addition, the essential infrastructure issues need to be assessed and established before the start of operations. These factors stimulated the preparation of the current publication for guiding the establishment of IRFs using a phased project milestones methodology and applying a tailored approach depending on the degree of complexity of the ionizing radiation facility.

This publication, developed under the IAEA regional technical cooperation project entitled Promoting Self-Reliance and Sustainability of National Nuclear Institutions, is intended to provide guidance for organizations and institutions working on ionizing radiation facility projects in order to enable them to undertake the project in a well organized manner. It includes some considerations for a feasibility study of an ionizing radiation facility that takes preliminary strategic planning, assessment of infrastructure issues and cost–benefit analysis as the key elements. In addition, it provides detailed methodologies for assessing the status of necessary infrastructure and determining the efforts required for such infrastructure development in a phased approach, and includes information on existing IAEA review missions and advisory services, databases and other electronic information resources. The guidance also aims to help Member States, as well as their respective organizations and institutions that are working on ionizing radiation facility projects, to understand their commitments and obligations associated with an ionizing radiation facility project prior to consideration of an ionizing radiation facility project; this includes an appropriate legal and regulatory framework for the safety and security of radioactive materials.

The IAEA wishes to acknowledge the assistance of all the experts, who contributed to the preparation of this publication, in particular N. Ramamoorthy (India) and I. C. Lim (Republic of Korea) who compiled, elaborated and reviewed the first complete version of the text. The IAEA officers responsible for this publication were D. Ridikas and N. Pessoa Barradas of the Division of Physical and Chemical Sciences.

CONTENTS

1. INTRODUCTION

1.1. BACKGROUND

The application of nuclear technology encompasses the use of nuclear energy and ionizing radiation resulting from nuclear reactions or radioactive decay. Nuclear and other ionizing radiations can be used in ways that benefit humankind. Such applications can provide benefits to the quality of life, health, environment, water security, food and agriculture, and industrial quality and productivity. Making use of radionuclides and related techniques can play a valuable role in managing medical needs, such as those related to the prevention, diagnosis and treatment of disease. Nuclear and isotopic techniques are used to probe and understand the world we live in, to provide decision makers with the information necessary to address modern environmental issues, to adapt to future scenarios and to mitigate and adapt to climate change. Water security — its availability, quality, management and protection — has become a critical issue in human development, as has environmental and economic sustainability, particularly in the light of global population growth. Nuclear technologies [1] provide competitive and often unique solutions to help fight hunger and malnutrition, improve environmental sustainability and ensure that food maintains its quality for longer, is safe and can be traded between different parts of the world. Nuclear science and technology play a major role in bringing innovation, safety and efficiency to industrial processes. They are also successfully applied for material analysis. Nowadays, material modification and its applications are expanding to the semiconductor industry and to the treatment and restoration of cultural heritage objects. The application of medical radioisotopes in the form of radiopharmaceuticals and sealed sources, as well as of photons, electrons and protons from accelerators, has vastly changed the status of human health and quality of life in patients with cancer, heart diseases and other ailments during recent decades. The fields of both diagnosis and therapy have been profoundly affected by the introduction of new generations of radiolabelled targeting molecules, produced by ionizing radiation facilities (IRFs) such as cyclotrons and research reactors.

Thus, countries use nuclear science and technology to contribute to their development objectives in areas including energy, human health, food production, water management, environmental protection, and climate change adaptation and mitigation. The use of this technology contributes directly to 9 of the 17 Sustainable Development Goals set out in the United Nations 2030 Agenda for Sustainable Development and is well in line with the IAEA's 'Atoms for Peace and Development' mandate [2].

Most of the above mentioned applications of nuclear science and technology emerge from the use of ionizing radiation. The devices and equipment that are the source of such radiation can be machines and/or radioisotopes. Whether from machines or radioisotopes, radiation sources are generally housed in purpose-built facilities to ensure that they are kept safe and that any radiation that they produce is shielded to prevent any undue exposure of the general population. Such facilities are collectively called IRFs. These IRFs include accelerator based neutron sources; gamma ray irradiation installations; cyclotrons and other accelerators for radioisotope production or other irradiation purposes; nuclear medicine centres for imaging and therapy, including electron LINACs; proton and hadron accelerators; R&D ion beam accelerator laboratories; radiopharmaceutical production facilities; analytical or dosimetry laboratories, etc.

The expansion of nuclear technology applications has driven an increase in the availability of IRFs, with many national nuclear institutions (NNIs), medical centres and universities as well as some commercial entities now owning and operating IRFs for their research, medical, service or industrial purposes. Furthermore, with time, the R&D activities of mature and potentially productive NNIs need to be adjusted for optimum sustainability and for the benefit of all parties concerned, since reliance on government or organizational funding needs to be balanced with other funding opportunities, including possible revenue generation on the basis of products and services. Similar arguments also hold for research and production facilities being an inseparable part of the valuable infrastructure of these NNIs, as they require significant investment and operational resources. The regional technical cooperation project entitled Promoting Self-Reliance and Sustainability of National Nuclear Institutions was designed to support the self-reliance and sustainability of NNIs through the development and use of the appropriate tools for developing strategic planning and infrastructure [3]. Consequently, self-reliance and sustainability are also key parameters to consider during the design, development and utilization phases of these IRFs.

With the exception of very small scale radiation equipment, the development of both 'hard' (facilities, equipment, building, etc.) and 'soft' (regulatory, training, quality management, etc.) infrastructure needs to be considered when deploying an IRF. This publication provides a structured approach to defining the scope and assessing the status of the relevant infrastructure issues through a set of well defined conditions for each one, which are provided in Annex I. At the same time, it is noted that several issues (nuclear safety, legal framework, safeguards, regulatory framework, environmental protection and nuclear security) are closely related to the national framework and are addressed by existing IAEA review missions and advisory services or by referring to existing IAEA databases and other electronic information resources. A list of these services and resources is given in Annex II.

Thus, the establishment of an IRF needs to be conducted through a well managed project, which increases the chances of successful implementation and full utilization once the facility starts operation and services.

1.2. RELEVANCE TO THE IAEA MILESTONES APPROACH

In the early 2000s, close to thirty countries were considering adding nuclear power to their national energy mix as a stable additional source of electricity. To help Member States in establishing the necessary nuclear infrastructure and to implement their nuclear power programme successfully, the IAEA developed the Milestones Approach, which is a phased, comprehensive method that enables a country to understand the commitments and obligations associated with developing a safe, secure and sustainable nuclear power programme [4]. Later, the same approach was adapted in the preparation of guidance for Member States on planning and implementing new research reactor programmes [5]. The use of the Milestones Approach for research reactor programmes arose from the fact that, in many ways, the complexity of the infrastructure issues associated with a new research reactor resembles that associated with a nuclear power reactor.

Many NNIs and other organizations — including those in the private sector — of IAEA Member States have deployed IRFs, and the interest of IAEA Member States in developing new research and production facilities involving nuclear and radiation technology has been growing. The establishment of an IRF requires the development of adequate infrastructure (or confirmation of the adequacy of an existing infrastructure), the implementation of which requires considerable resources to ensure the safe and secure construction, operation and effective utilization of the IRF throughout its lifetime as well as safe management of radioactive waste (where applicable) followed by decommissioning. These characteristics ensure that, in principle, some of the elements of the Milestones Approach described above for a nuclear power plant or research reactor can be also applied in the case of IRFs.

Nevertheless, there is an important difference that needs to be considered. While the Milestones Approach is directed at the programme level, the establishment of an IRF, although it can be implemented within a broader national programme (examples include to reduce the dependency of a country on radioisotope imports, support exports of agricultural produce, and expand national capability in nuclear sciences and technologies, among others), requires a project approach.

Projects usually move through three phases: pre-project, project formulation and project implementation. In phase 1, the pre-project phase, consideration is given to whether the project is justified; preliminary plans and infrastructure assessments are made, and a cost–benefit analysis is undertaken before a formal decision to proceed and develop a facility is taken. The project formulation is phase 2; once the decision is taken to progress to the building of an IRF, this second phase of the project involves the carrying out of preparatory work and the development of a detailed specification for the IRF. In phase 3, project implementation, the IRF is built according to the specifications.

Milestones are a project management tool commonly used to mark specific checkpoints along the timeline of a project as it moves through the different project phases. These milestones are points that may signal 'anchors', such as a project start and end date, and also flag the attainment of significant developments and a need for external review or input and budget checks. In many instances, milestones define progress in implementing the project rather than project duration. They focus on the major progress points as a series of sequential steps that need to be reached to achieve overall project success [6]. The use of project milestones in a phased project approach allows the project team and other project stakeholders to view the current status of a project, identify deviations and detect errors in the data collection method of progress reports. In addition, milestones are an excellent tool to communicate the status of the project to stakeholders external to the project team [7]. Therefore, it is expected that the use of a phased approach for project management will be helpful for many NNIs and other organizations in deploying their IRFs, as it contributes to strengthening their sustainability and increasing their self-reliance.

1.3. OBJECTIVES

This publication is intended to assist national public or private organizations that are considering the establishment, or are undertaking a major upgrade or refurbishment, of an IRF to achieve the following:

— Make an informed judgement of the necessity for the project and assess the organization's readiness to initiate such an undertaking, taking into account the existing gaps in the national and/or organizational infrastructure and considering the required commitment during the entire lifetime of the IRF, including siting, design, construction, commissioning, operation and decommissioning and all operational activities;

— Determine the magnitude of the commitment necessary to establish and achieve the effective and efficient use of an IRF in a safe, secure and technically sound manner;
— Conduct a feasibility study to provide objective information to decision makers and other stakeholders about the utility and soundness of the IRF project;
— Support the success and sustainability of an IRF and limit the risk of future underutilization or legacy issues with the facility;
— Allow for a timely decision not to pursue a new IRF or refurbish an existing one (if the utility and sustainability of the facility are shown to be insufficient by the feasibility study).

This publication provides guidance to the organization on the sound justification[1] of this type of project, such that it can confirm that it has undertaken the following:

— Justified the need for an IRF within a well identified organizational or national strategy;
— Comprehensively recognized and identified the commitments and obligations associated with the establishment of an IRF;
— Established and adequately prepared the infrastructure prerequisite to the establishment of an IRF;
— Established all the competencies and capabilities necessary to operate an IRF safely and securely, in line with the IAEA safety standards and nuclear security recommendations, and economically over its lifetime, and to manage the ensuing radioactive waste;
— Established adequate funding and financing mechanisms for the IRF project throughout its life cycle, including for the decommissioning phase [8].

1.4. SCOPE

This publication is directed at facilities of a dimension or complexity sufficiently large to warrant application of the phased approach to project management described here. Small scale mobile or portable facilities, such as nucleonic gauges or hand-held X ray fluorescence analysers, are not

---

[1] In this publication, the term 'justification' is used in its generic and everyday sense (i.e. that there is a need for an IRF based on the organizational, national or regional needs for its services, the availability of alternatives, and the availability of sufficient financial, technical and human resources) and not as a safety oriented principle.

within the scope of this publication. Nevertheless, the types of IRF targeted by this publication are numerous and diverse. These IRFs employ a variety of technologies, radiation sources, particles and resulting ionizing radiation quantified by its dose rates. Some typical examples are provided below, grouped into three main categories:

- R&D and related facilities:
    - Ion beam accelerators;
    - Synchrotron light facilities;
    - Laboratory based experimental facilities using isotopic neutron sources (Cf, Am-Be, Pu-Be, etc.) or DD/DT neutron generators;
    - Neutron source facilities based on particle accelerators;
    - Experimental controlled fusion facilities or plasma devices.
- Medical service oriented and related facilities:
    - Cancer treatment facilities based on electron linear accelerators (LINAC) or on radioactive sources;
    - Cancer treatment facilities based on proton or heavy ion accelerators;
    - Medical cyclotrons and associated radiopharmacy activities (some may also have a positron emission tomography/computed tomography (PET/CT) imaging facility);
    - Nuclear medicine centre for diagnostic imaging (single photon emission computed tomography/computed tomography (SPECT/CT) or PET/CT) and therapy;
    - Radiopharmaceutical production/radiopharmacy facilities.
- Industrial/commercial service oriented and related facilities:
    - Calibration and dosimetry services, based on portable or laboratory stationed equipment;
    - Radiation based (as a probe) industrial process/system diagnostics services;
    - Laboratory based radiography service for non-destructive testing;
    - X ray/electron beam, gamma irradiation service centre for radiation processing applications.

The IRFs that are based on the use of radioactive sources of high risk type — Categories 1 and 2 under the IAEA classification as indicated in IAEA Safety Standards Series No. RS-G-1.9, Categorization of Radioactive Sources [9] — have been of increasing concern for many authorities (such as national regulatory authorities) dealing with the security of radioactive materials. This is due to the potential vulnerability to the sabotage or theft of high risk sources for malicious purposes (radiological terrorism). Examples include the radioisotope sources and associated systems used in cancer therapy, gamma radiation plants for the sterilization of medical products or food preservation, blood research irradiator units and industrial radiography devices. There is hence advocacy to adopt alternative technologies to the use of high risk radioactive sources [10–12] wherever possible, despite known technological and financial challenges in many of these cases, especially for wider deployment in different regions of the world.

The IAEA guidance and support available to its Member States in planning and/or establishing IRFs, including in ensuring radiation safety (IAEA Safety Standards Series No. GSR Part 1 (Rev. 1), Governmental, Legal and Regulatory Framework for Safety [13], and GSR Part 3, Radiation Protection and Safety of Radioactive Sources: International Basic Standards [14]) and the security of high risk radioactive sources [15–18], will therefore continue to be of high importance. There has also been support through global cooperation initiatives for fostering the adoption of alternative technologies in select cases (e.g. electron accelerators for external beam radiotherapy of cancer, X ray based blood irradiators). However, currently no satisfactory alternative is available for some vital applications (e.g. brachytherapy for certain cancers), while development of alternative technologies is continuing in some other areas (e.g. enhancing the ruggedness and reliability of electron accelerators for sustained continuous operation, portable X ray based systems for radiography in remote areas or open field conditions). A discussion of the alternative technologies to radiation sources is beyond the scope of the present publication. The brief narration above is meant to inform the stakeholders and other users of the publication about this emerging topical issue, which is likely to impact certain types of IRF projects in future.

The scope of this publication covers the establishment of new IRFs and the refurbishment or upgrade of an existing one with input and support from multiple stakeholders. It also presents a phased approach framework (milestones) in the development of necessary infrastructure. Emphasis is given to phase 1, 'pre-project', through the presentation of a methodology to identify gaps in the infrastructure, which are addressed in the subsequent project phases. The publication addresses the entire lifetime of the IRF holistically, including siting, design, construction, commissioning, operation and decommissioning, and all operational activities including ageing management.

The information presented in this publication is based on the experience and good practices of countries which have recently established IRFs and is not intended to impose new or absolute standards on those contemplating an IRF.

1.5. STRUCTURE

This publication consists of three main sections in addition to the introduction: Section 2 describes the project phases and lists necessary infrastructure issues and milestones for a new IRF project. It also introduces the graded approach[2] for nuclear safety and security and a tailored approach, which has to be individually developed and tailored for specific IRFs, considering the varieties among IRFs and their differences compared with nuclear facilities.[3] Section 3 provides guidelines for the feasibility study of an IRF, which includes the preparation of a preliminary strategic plan (PSP), performing infrastructure assessment and cost–benefit analysis. Section 4 is an executive summary, emphasizing the importance of a well structured holistic approach, organized in distinctive phases and with specific milestones for the establishment of a new IRF or refurbishment of an existing one.

Annex I accompanies Section 3 and contains tables with infrastructure conditions that are based on the phased Milestones Approach and specifically adapted for IRF projects with the application of a tailored approach and in consideration of the differences between programme and projects levels.

Annex II provides a list of relevant IAEA review missions and advisory services, databases and other electronic information resources.

1.6. USERS — TARGET AUDIENCE

All stakeholders involved in the development, regulation, operation and utilization of an IRF are targeted users of this publication. However, their major interests may vary and depend on how they are engaged in a new IRF project.

NNIs or private sector entities planning to establish an IRF may use this publication for reassurance or justification of the need for a facility and to conduct a comprehensive feasibility study. The publication will be also helpful for identifying and addressing infrastructure issues related to a specific IRF.

Policy and decision makers, including development planning and funding entities, can use this publication to gain insights into the overall process of what needs to be done for the planning, installation, operation and use of the IRF. The

---

[2] In general, a 'graded approach' means a structured method by which the stringency of control to be applied to a product or process is commensurate with the risk associated with a loss of control.

[3] The IAEA Nuclear Safety and Security Glossary [19] defines 'nuclear facility' as a facility (including associated buildings and equipment) in which nuclear material is produced, processed, used, handled, stored or disposed of.

contents of this publication may also be used to check whether all these steps are well prepared. It also provides valuable information on the cost–benefit analysis for a new IRF.

Authorities regulating the IRF (e.g. regulatory bodies for radiation safety) may use this publication as a tool to assist them in identifying gaps or areas that need strengthening in the legislative and regulatory frameworks and other relevant infrastructure issues having to do with the regulatory oversight of IRFs.

For the users of an IRF, this publication will help them to recognize that their needs and requirements are sufficiently covered and to determine if they need to remain engaged in the planning and other establishment processes in order to benefit from the full use of the IRF later on.

The IAEA may also use this publication to assist Member States that are planning or are already establishing new IRFs so that the Agency's assistance can be provided in a coordinated, effective and timely manner.

Finally, and to a limited extent, this publication might also be beneficial to the technology providers and vendors to assist less experienced organizations in the establishment of new IRFs in a well structured and phased approach.

1.7. FRAMEWORK FOR SAFETY AND SECURITY

1.7.1. Framework for safety and security

An appropriate national legislative and regulatory framework for safety and security, in line with IAEA standards for radiation safety and IAEA recommendations for security of radioactive material, needs to be in place to provide for the protection of the patients, the public and workers. It is a prerequisite to considering an IRF project. Member States are further advised to make appropriate use of the IAEA safety standards and guidance in the IAEA Nuclear Security Series relevant to radiation sources, including those relating to the governmental, legal and regulatory framework for safety and security.

In this regard, the international framework for safety includes relevant international legal instruments and the IAEA safety standards. The IAEA safety standards reflect an international consensus on what constitutes a high level of safety for protecting people and the environment from the harmful effects of ionizing radiation. The IAEA Safety Fundamentals present the fundamental safety objectives and principles of protection and safety and provide the basis for the safety requirements. IAEA safety requirements provide the basics of the regulatory infrastructure and can support an IRF to identify the applicable requirements to be met and measures to be taken in order to ensure the protection of people and the environment, both at present and in the future. IAEA Safety

Guides provide recommendations and guidance on how to comply with the requirements. Specific Safety Guides are available for different IRFs.

The international framework for security includes relevant international legal instruments and the IAEA Nuclear Security Series, which provides international consensus guidance on all aspects of nuclear security to support States as they work to fulfil their responsibility for nuclear security. IAEA Security Fundamentals provide the basis for a security related programme. IAEA Nuclear Security Recommendations set out measures that States should take in order to achieve and maintain an effective nuclear security regime. IAEA Nuclear Security Implementing Guides provide guidance on how States can implement the recommendations. IAEA nuclear security Technical Guidance provides more detailed guidance on specific methodologies and techniques for implementing nuclear security measures.

1.7.2. Code of Conduct on the Safety and Security of Radioactive Sources

The Code of Conduct on the Safety and Security of Radioactive Sources (Code of Conduct) and its Supplementary Guidance [20] contain non-legally binding principles and guidance for the development and harmonization of national polices, legislation and regulations on the safety and security of radioactive sources, on responsibilities of the regulatory body, on the import and export of radioactive sources and on the management of disused radioactive sources. A purpose of the Code of Conduct and its Supplementary Guidance is to assist Member States in ensuring that radioactive sources are used within an appropriate legislative and regulatory framework for the safety and security of radioactive sources. It provides 'best practice' guidance to the State, the regulatory body, the end users and some technical support organizations for the management of the safety and security of radioactive sources.

The Code of Conduct is based on IAEA safety standards and nuclear security guidance, and Member States are encouraged to make a political commitment to the Code and its Supplementary Guidance on the Import and Export of Radioactive Sources and on the Management of Disused Radioactive Sources and to work towards implementing their provisions. National implementation of the Code of Conduct provisions may be achieved through the development or amendment of national safety and security legislation and regulations. In accordance with resolution GC (48)/RES/10 on Measures to Strengthen International Cooperation in Nuclear, Radiation and Transport Safety and Waste Management [21], Member States are encouraged to use the Code of Conduct as the basis upon which to regulate and conduct activities involving radioactive sources.

Member States that are considering an IRF project are encouraged to implement the provisions of the Code of Conduct and its Supplementary Guidance. In line with establishing a national infrastructure for safety and security, such implementation needs to be accomplished prior to the start of the IRF project through existing or amended national safety and security legislation and regulations pertaining to all stages in the life cycle of radioactive sources, including the management of the disused radioactive sources.

2. IRF PROJECT PHASES AND MILESTONES

The main factors and considerations in undertaking a project to establish a new IRF are listed below:

— In most cases, establishment of an IRF will require multiple stakeholders to be consulted and involved, including those outside of the organization that is planning, leading and implementing the project.
— In the planning of an IRF, it is crucial to develop a sound justification of the need for the IRF and a sustainable utilization strategy that considers the opinions of multiple stakeholders and users (or customers/clients in the case of commercially oriented IRFs).
— The establishment of an IRF requires adequate infrastructure to be in place as well as the investment of substantial resources to cover associated costs for the life of the facility, including those for development of human resources. Therefore, a feasibility study is necessary to determine whether to implement such a project, and what the associated costs will be.
— In most cases, the entities that conduct the IRF project would not have the capacity to develop the IRF by themselves. Therefore, a technology (or a major equipment) supplier is usually selected through a bidding process, and the preparation of bidding documentation[4] is an important activity.

---

[4] If the organization decides to work with a single technology provider without a competitive bidding process, documentation of functional and technical specifications of a new IRF will still be needed for contractual purposes.

2.1. SYNERGIES AND DIFFERENCES WITH THE MILESTONES APPROACH FOR RESEARCH REACTORS

Research reactors are nuclear installations, while IRFs are radiation facilities where there is usually no nuclear material produced, processed, used, handled, stored or disposed of. Nevertheless, there is a similarity between them when it comes to their use. Research reactors are scientific and technological tools used to produce radioisotopes or other irradiation products; perform research on the analytical characterization of materials and objects using neutrons originating from fission reactions; and provide education and training in nuclear sciences and technologies. Similarly, an IRF is a tool used to change or analyse the properties of materials and substances or to produce radioisotopes using radiation, or it is directly used for medical purposes to treat or diagnose diseases. Both research reactors and IRFs provide services or products to a variety of users such as universities, research institutes, medical centres and industries. From these, it is readily apparent that the broad uses of IRFs and research reactors are similar. By the same token, both research reactors and IRFs, together with associated activities, pose radiation risks to workers and the public and to the environment, so safety has to be ensured, including through a governmental, legal and regulatory framework for safety and security and through appropriate safety management by operating organizations and regulatory control by mandated authorities.

The IAEA publication Specific Considerations and Milestones for a Research Reactor Project [5], also known as the Research Reactor Milestones publication, includes a detailed description of 19 infrastructure issues that need to be addressed and the expected level of achievement (or milestones) at the end of each phase of the project. It is proposed that a similar holistic methodology be applied for the planning and establishment of IRFs, employing a graded approach and tailoring it as appropriate to the nature and complexity of the IRF. In addition, the IAEA has developed a number of other detailed and practical guides for associated topics related to project management and implementation relevant to the establishment of a new research reactor that can also be applied, with some adjustment and gradation, to the project management of new IRFs:

— Feasibility study [22];
— Development of strategic plan [23];
— Specific considerations in the assessment of infrastructure issues [24];
— Preparation of bid specification [25].

When establishing a new IRF, it is vital to assess the status of the relevant infrastructure issues, determine the work needed to develop the infrastructure to the necessary level and systematically ensure its readiness, in parallel with the

as targets or shielding components (e.g. for the production of radioisotopes by fission or for research purposes), there will be more considerations for security aspects, radiation safety management and also possibly safeguards.

2.3. CONSIDERATIONS FOR A TAILORED APPROACH

IRFs can be managed and operated by NNIs, universities, medical centres or private entities or companies. In addition, their functions and users may vary extensively depending on the facility's specifications and purpose. For example, the main users of an electron LINAC are different from those of a cyclotron producing radioisotopes or an electrostatic ion beam accelerator.

As described above and in Sections 1.3 and 1.4, this publication covers a wide range of facilities and activities other than process control or regulation, including their distinct differences as compared with nuclear facilities. Therefore, it might be practical to consider a tailored approach to using this publication more broadly (not limited to the safety and security aspects covered by the graded approach described above) and take into consideration specific aspects of different facilities and associated activities practiced in these facilities.

Some attributes for the broader application of such a tailored approach can be illustrated through the examples described below:

— *Facility owner/operator: public organizations vs. private entities.* In most cases, nuclear research institutions belong and report to national governments. Therefore, if users of this publication are from government funded organizations, the publication has to be used with an emphasis on public services and the creation and dissemination of scientific knowledge and products. Private entities, such as private hospitals or private industry, would likely also be interested in generating benefits from providing services or goods through a more effective and efficient use of IRFs.

— *Users: national facilities vs. international facilities.* In some instances, a facility may be installed for use by a single entity or institution. In other cases, an IRF may be established to serve a large number of users. These users may be limited to domestic entities or could include users from other countries. As a result, different management, organizational and access procedures for national and international users/customers might apply.

— *Types of service provided: analytical services, healthcare, irradiation services and products.* The services provided by the IRF may be broadly grouped into analytical services, healthcare and irradiation services and products. There will be differences in technical standards, quality assurance/quality control (QA/QC) requirements or in regulations (other than nuclear safety and nuclear security related, e.g. drug master file in the case of production of radiopharmaceuticals) that need to be satisfied for each case, and considerations for efficient and effective service provision.

— *Newly built vs. replacement or refurbishment.* In some cases, the IRF to be established will be a new installation, and in other cases, it is to replace or refurbish an existing IRF because of its ageing, maintenance or utilization issues. There may be many more aspects to be considered with regard to a new facility. In the latter case, it will also be important to reflect on the lessons learned from operating the existing IRF, new needs and possible changes in national regulations.

2.4. PROJECT DEVELOPMENT OF A NEW IRF

As mentioned in Section 2.1, a project milestones methodology can be adjusted and applied to suit IRFs. The three project phases of development and their corresponding milestones are described in Table 1 and shown schematically in Fig. 1, as adapted from Ref. [5] by focusing on the project development aspects.

With regard to IRFs, the various 'soft' and 'hard' infrastructure issues can be categorized according to 19 generic headings (Table 2). It needs to be emphasized that the order of these 19 issues as presented in Table 2 does not indicate a rank of importance or hierarchy. Each issue is equally significant (although its importance can vary, depending on the nature of the specific IRF in question) and requires careful consideration. Each of the issues will need to

TABLE 1. PROJECT DEVELOPMENT PHASES AND MILESTONES
(adapted from Ref. [5])

Phase	Description	Milestone
(1) Pre-project	Justification of the need for an IRF and considerations before a decision to launch an IRF project is taken	Decision is taken for establishment of an IRF
(2) Project formulation	Preparatory work for the establishment of an IRF after a strategic decision was taken	Request for proposal for the IRF is issued
(3) Implementation	Activities to design and construct an IRF	Licence/permit is received

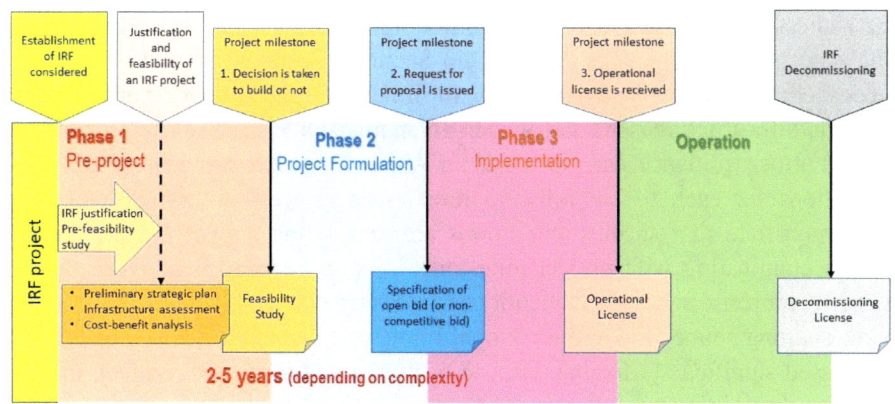

FIG. 1. Phases of an IRF development project (adapted from Ref. [5]). In some countries a pre-operational (construction/testing) licence is required, depending on the type of IRF.

TABLE 2. INFRASTRUCTURE ISSUES AND PROJECT PHASES
(adapted from Ref. [5])

Issues		Phase 1: pre-project	Phase 2: formulation	Phase 3: implementation
1.	Organizational position			
2.	Nuclear and radiation safety			
3.	Management			
4.	Funding and financing			
5.	Legal framework	CONDITIONS: IDENTIFYING GAPS	CONDITIONS: CLOSING GAPS	CONDITIONS: CLOSING GAPS
6.	Safeguards			
7.	Regulatory framework			
8.	Radiation protection			
9.	Utilization			
10.	Human resources development			
11.	Stakeholder involvement			
12.	Site survey, selection and evaluation			
13.	Environmental protection			
14.	Emergency preparedness and response			
15.	Nuclear security			
16.	Fostering expansion of ownership			
17.	Radioactive waste management			
18.	Industrial involvement			
19.	Procurement			

be addressed and assessed in a feasibility study, as they all have a bearing on the success of a project. As discussed in Section 2.3, above, a graded approach needs to be applied while assessing the fulfilment of different conditions for each specific infrastructure issue and at each given milestone.

Further guidance can be found in Annex I, which provides tables with conditions for each of the infrastructure issues as well as examples of how to demonstrate that specific conditions are met at the end of the pre-project phase, culminating with project milestone 1 (i.e. a decision is taken to launch the IRF project (or not)). As mentioned in Section 2.1, a number of the issues have a strong connection to the national framework, which is not expected to be addressed significantly for the purpose of implementing an IRF project, mainly because those infrastructure issues are expected to already be in place in most of the countries. These issues are nuclear and radiation safety, legal framework, safeguards, regulatory framework, environmental protection (particularly in those aspects that concern the national framework) and nuclear security.

A precondition for developing a successful IRF project is that no major gaps in the infrastructure development for those issues exist, as the national framework is not expected to require significant changes for the purpose of establishing an IRF. For those issues, the conditions in Annex I may be met in a simplified way, namely by referring to the results of IAEA review missions and advisory services [27] or by referring to existing IAEA databases and other electronic information resources [28]. A list of these services and resources is given in Annex II.

2.4.1. Pre-project

During project phase 1, a public organization or private entity has to complete the justification for the need of an IRF project, which delivers project milestone 1 with the production of the feasibility study report (FSR). The FSR demonstrates that the organization is in a position to make an informed decision as to whether to proceed with the IRF project.[5] All the activities required to reach the project milestone 1 are directed by a project management team (PMT) and conducted by a project implementation team[6] (PIT). The role of the PMT is (i) to review and accept, as appropriate, the work done and recommendations provided

---

[5] If the IRF proposed involves the use of radioactive sources of Category 1 or 2 (desirable for Category 3, too), careful assessment of the scope of the use of alternative technologies also has to be performed and the findings considered in the FSR.

[6] In cases where a small IRF is established, requiring quite limited infrastructure or with most of such infrastructure already in place, the PMT's functions are often performed by the institution's Head.

by the PIT, and (ii) to ensure that the necessary infrastructure and resources will be in place prior to the establishment of the IRF.[7] The PIT is formed to study, develop and promote the IRF project by conducting a comprehensive feasibility study. Its activities include but are not limited to: formulating the justification of the need for the IRF, developing its functional and technical specifications, and recommending actions that need to be taken to reinforce or implement the required infrastructure and address the policy issues. As was already mentioned in Section 1.5, the FSR consists of three pillars: developing a PSP; implementing infrastructure assessment; and performing a cost–benefit analysis.

The PSP is used to help gather input and support from the potential stakeholders, users and suppliers (including international support, if applicable or necessary), as well as to provide clear guidance to decision makers on the actions expected from them for the continuation of a safe and successful IRF project. The PSP will summarize the justification of the need for the IRF and its associated facilities, as well as their functional and technical specifications, and will develop detailed recommendations for an organizational (management) structure of the IRF, including required resources. The PSP will also include the strategy of planning the IRF project, if necessary and appropriate, as well as the policy decisions and actions required from the PMT. It also communicates the necessity of ensuring the safety, security and peaceful use of radiation sources for those IRFs where it is applicable. The PSP will be updated and enhanced regularly during the establishment phases of IRF so that later it evolves into a strategic plan for the utilization of an IRF. It may be necessary to consider possible modifications in research areas or products and services this facility is tasked to provide. In some cases, the purpose or scope of the IRF might also change with time, and therefore the facility's strategic plan will need to be revised or expanded accordingly.

Before embarking on the IRF project, the public or private organization has to develop a comprehensive understanding of the obligations and commitments involved and ensure that there is a clear strategy and resources available to discharge them. At milestone 1 of a new IRF project, decision makers and other relevant stakeholders will expect confirmation that the IRF project is supported by a comprehensive review of infrastructure necessary for the specific IRF and that all possible gaps have been identified and documented in the infrastructure assessment report (IAR) as part of the FSR.

The infrastructure required to support an IRF project varies widely and includes the physical facilities and equipment associated with the IRF and related

---

[7] For the infrastructure issues that have a strong connection to the national framework, the PMT's role is either to verify that they are already in place or to ensure that any remaining gaps are closed.

infrastructure (e.g. if applicable, a radioactive waste management facility). If radioactive or nuclear materials will be used in such facilities, their supply chain and logistics along with associated nuclear security and import/export control aspects need to be considered as well. It will require sufficient human and financial resources throughout its life cycle. In some cases, revision of the legislative and regulatory framework might be needed. This publication does not directly address such revision, which is usually undertaken under a national programme developed for that purpose.

Cost–benefit analysis is an important part of the feasibility study, whether the IRF is built to achieve R&D goals or provide products and services for social or economic benefit. The estimated costs for design and construction of the IRF as well as resources needed for adequate infrastructure development will be a major part of the capital costs. In addition, the lifetime costs for operation, maintenance, upgrades and decommissioning have to be included in the overall cost estimate.[8] In addition to cost analysis, the evaluation of the benefits from the operation of an IRF needs to be available. The benefits may be tangible or intangible depending on the characteristics of the services provided by the IRF. Their use for the benefit analysis will depend on the analysis method defined by the PMT or the authority who will make the decision about the IRF project.

In most cases, the establishment of management systems is of key importance for the success of the IRFs, and they already need to be implemented during the pre-project phase. For example, management systems have proven to be a fundamental tool to support the fulfilment of quality, safety and functionality expectations. Management systems provide IRFs with the ability to achieve the highest levels of compliance with customer and applicable legal and regulatory requirements as well as customer satisfaction. They are also instrumental in promoting the safety and security culture, whereas environmental management systems and occupational health and safety management systems contribute highly to the overall results. Some of these results are intangible aspects directly related to the human factor: for instance, while an improvement in public perception due to a good organization's environmental and safety performance increases the acceptance and sustainability of radiation processing, at the same time the workers' motivation and sense of belonging are enhanced by the increase in their confidence and trust that their health and safety are just as important to their organization's authorities as quality and productivity.

---

[8] In the case of IRF projects involving the use of radioactive sources of Category 1 or 2 (desirable for Category 3, too), the cost of end-of-life management of the sources (along with an operational strategy for the same) should be also taken into consideration.

2.4.2. Project formulation

Following the strategic decision to proceed with the development of an IRF project, substantive work needs to be undertaken to achieve the necessary level of technical and institutional competence. During the second phase of the project, the facility owner or operator organization will carry out the work required to prepare for the construction of an IRF and the preparation of the bid specifications[9] (or selecting a single technology provider), which is the milestone for project phase 2. The infrastructure gaps identified during the pre-project phase are sufficiently addressed to the point of readiness to request a bid or enter into a commercial contract. Before the commencement of the bidding process, the licensing stages and activities subject to licence need to be defined, including safety and security requirements for the bidding process itself, along with the bid evaluation. An effective project management system and staff capabilities need to be adjusted or developed to ensure proper accomplishment of the obligations of the entity that will own and operate the IRF (or both entities if the owner and operating organizations are different). The operating organization has a key role at this time in ensuring that it has developed the competencies to manage an IRF project, to achieve the level of organization and safety and security culture necessary to meet the regulatory requirements, and to demonstrate that it is an adequately informed, duly resourced and effective organization [32].

2.4.3. Project implementation

After a technology provider (vendor) has been selected, the third phase of the project development consists of all the activities necessary to build the IRF and close any remaining gaps in the necessary infrastructure development. The milestone for project phase 3 will be the acquisition of an operating licence/permit.[10] During this phase, the largest capital expenditures will occur, and the involvement of all relevant organizations and stakeholders is crucial to the successful outcome, as all have important roles to play. At the end of this phase, the operating organization will have evolved from an organization conducting an IRF establishment project to an organization that, in addition to the responsibility for siting, design and construction, can accept responsibility for commissioning, operating and eventual decommissioning of the facility.

---

[9] In some cases, a 'turnkey' solution for IRFs might not be possible, and therefore the entire project might require a number of bidding processes for the main equipment, auxiliary systems and components, building construction, etc.

[10] Other licences and permits might be needed at previous stages, as well. The operating licence is the milestone for project phase 3.

Procedures and arrangements need to be developed to ensure the safe and secure management of the IRF under all conditions, fulfilling the national requirements and in line with the international safety standards, as well as for adequate human resource development and training for all necessary levels of staff.

While achieving the third milestone is a major accomplishment, it is also necessary to recognize that, at this moment, commitment to safety has to be reinforced and considered as a long-lasting commitment to the safe, secure and effective lifelong operation and utilization (and ulterior decommissioning) of the IRF. Periodic safety assessments with a graded approach and peer reviews in areas such as safety, operation and maintenance have a positive impact on the safety and performance of the IRF.

2.5. RISK MANAGEMENT

Risk can be defined as the effect of uncertainty on an objective, whether that be a positive opportunity or a negative threat; in this sense, risk does not always have a negative outcome. Risk management is the identification, evaluation and prioritization of risks followed by the coordinated and economical application of resources to minimize, monitor and control the probability or impact of potential events or to maximize the realization of opportunities. Because it cannot be entirely avoided, risk is a major factor to be considered during the management of an IRF project, and an appropriate and robust evaluation of risk needs to be an integral part of strategic decision making. Some amount of risk taking may be inevitable if a project is to achieve its objectives, and risks need to be identified and managed as far as possible.

Risk management is a vital component of the good management practices to be undertaken throughout the lifetime of an IRF, including during the feasibility study preparation. It is also an essential part of all IRF projects, irrespective of facility type, ownership, users and so on. Project management needs to control and contain risks if an IRF project is to stand a chance of being successful. This is because any project takes place in an uncertain world in which the future cannot be predicted with certainty. Documenting the PMT's approach to risk management and demonstrating that risks are identified and have to be controlled is an important element of the FSR, enhancing the viability of the project.

The task of risk management is to manage a project's exposure to risk, that is, to assess the probability of specific risks occurring, to reduce it where possible and to mitigate the potential impact on the project if the risk did occur. The aim is to manage that exposure by taking action to keep it at an acceptable level in a cost effective way. A risk management plan and a risk register are widely used as a means of documenting perceived risks and assessing their importance, and for recording actions that are to be, or have been, taken to manage those risks. A risk register can be a very simple document, but is a powerful means of communication when information is to be shared between all parties involved in the identification and management of risks. A typical risk register might contain the following:

— All the hazards related to the IRF project;
— The risks identified as resulting from those hazards;
— The estimated degree of risk from each (likelihood multiplied by impact);
— The planned response;
— At what stage of the project the response will be given, and by whom;
— The estimated effect of the response;
— Who will carry the consequence of the risk, should it occur.

In an IRF project, the actions of external stakeholders over whom the PMT may have little or no direct influence, such as the regulators and customers, may significantly impact the implementation of an IRF project. Therefore, the PMT will need to develop a strategy for seeking to communicate with and influence these and other external stakeholders to reduce the likelihood and impact of these risks as part of a stakeholder engagement strategy. Some of the external risks that could arise for the IRF PMT may include issues such as the following:

— Uncertainties regarding external supports, including government commitment and support;
— Disagreement regarding the assessment of potential user and customer needs and how to respond to them in terms of the nature of the products and services and their volume;
— Uncertainties or significant changes to the functional specifications of the IRF facility;
— Delays by external stakeholders in taking decisions that are critical to the implementation of the project;
— Difficulties in obtaining authorizations from the regulators to proceed with key stages of the project;
— Delays or disruptions to the preliminary siting process resulting from significant opposition to the project from lobby groups.

It would be normal practice for risk owners to be required to review the risks assigned to them and for the senior management of the project to review the top level risks at regular intervals, depending on the characteristics of an IRF project. More details on the methodology of risk analysis, assessment and management specific to a research reactor project, including some illustrative examples, can be found in section 3.5 of Ref. [19] as well as in annex II to Ref. [20].

3. CONSIDERATIONS FOR A FEASIBILITY STUDY

Phase 1 activities will conclude with a comprehensive FSR as the achievement of project milestone 1. The objective of the FSR is to provide comprehensive information necessary for the responsible authority or entity to make a decision on whether to proceed with the IRF project.[11] As already mentioned in Section 2.1, this publication adopts the methodology where the three pillars of the FSR are (i) a PSP, (ii) assessment of necessary infrastructure and (iii) cost–benefit analysis. This section describes the key points that can be applied to the case of the establishment of the IRF with regard to the three elements of the feasibility study, with the major steps and key elements outlined in Fig. 2.

The PIT needs to include individuals with high technical credibility and broad know-how. For example, they need to be highly knowledgeable in the technical fields of the IRF under consideration, have experience in financial and marketing aspects, and have good contacts with key stakeholders and users. The majority of the PIT members may be employees of the main organization supporting the IRF project, but this pool of experts might also be expanded, particularly if the required expertise is not available internally. The PIT would be responsible for identifying the stakeholder groups and their needs as well as for the development of quantitative requirements that meet their needs. These requirements would result in drafting functional specifications for the IRF and associated facilities and ultimately developing a conceptual design for the IRF itself.

3.1. PRELIMINARY STRATEGIC PLAN

3.1.1. Objectives

The objectives of the PSP for an IRF are twofold: to communicate to stakeholders and users the goals of the organization planning the IRF and to set out the actions needed to achieve those goals within the organization. The key reason for preparing a PSP is that it will provide an objective justification of

---

[11] The FSR could also provide, wherever applicable, adequate information on the choice of the particular technology proposed for the IRF after taking into account any available alternative technologies to the use of radioactive sources.

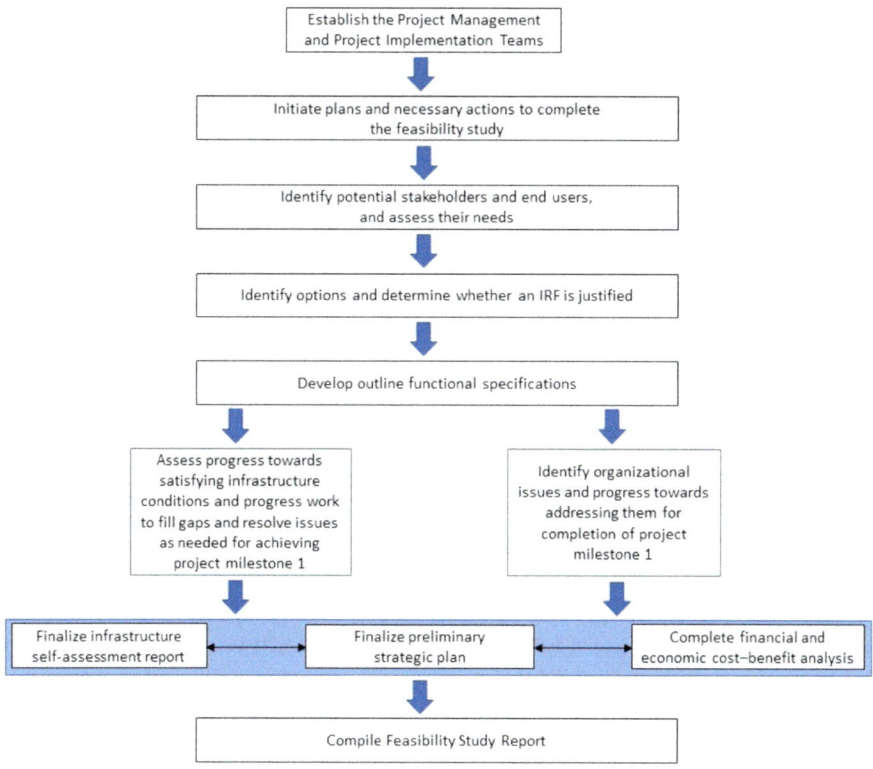

FIG. 2. Major steps for completion of a feasibility study (derived from Ref. [22]).

the need for a new IRF.[12] Thus, it has to identify key stakeholders and users, including present and future customers or clients, and provide confirmation of their existing or envisaged needs in terms of expected products and services. This is a key component in the justification of the need for the IRF and it can be used as a basis for the development of the functional specifications of the IRF and its associated facilities, as may be necessary.

It should be noted that, in addition to Ref. [23], the IAEA has also developed an e-learning course on Strategic Planning for National Nuclear Institutions [35]. This course provides introductory guidelines and methodologies for the development of a strategic plan for the efficient and sustainable utilization of an existing or planned IRF, illustrated by some representative examples.

---

[12] A similar approach can be used when planning major refurbishment, modification or upgrades of an already existing IRF. Equally, a sound strategic plan will be needed to ensure enhanced and sustainable utilization of an existing IRF.

3.1.2. Methodology for developing a preliminary strategic plan

The process of creating a PSP mainly involves applying analytical common sense, and some sections of the PSP (if applicable) might require documenting what has been conducted by the same already existing or similar type facilities, including lessons learned and good practices from there. As schematically indicated in Fig. 3 (derived from Ref. [23]), the PSP addresses how to develop and ensure a high level of utilization of the IRF by asking two simple questions: 'What should I do now?', followed by 'What can I do now?'

In the case of a new IRF, the emphasis in the PSP is to first determine 'what should I do now?' on the basis of actual and future stakeholder/user needs, as indicated by the arrows pulling towards the left in Fig. 3. Once this is understood, 'what can I do now?' will be progressively developed in assessment exercises, as indicated by the arrows pulling towards the right in Fig. 3. The expansion of the facility status circle to the right reflects more effectively defining/expanding its capabilities (i.e. can do) to meet more user needs; the expansion of the current stakeholder requirement circle to the left indicates more effectively covering/expanding the needs of stakeholders (i.e. should do) in order to make the circles overlap to the maximum. It is the responsibility of the IRF management to identify the intersection (shaded area) and to expand on this by matching and improving the abilities of the facility by actual and future stakeholder requirements. Such experiences can then also possibly provide good practices and lessons learned to other similar facilities.

An additional advantage of developing the PSP is to document the logic and analysis behind the facility's planned utilization. In the case of a new

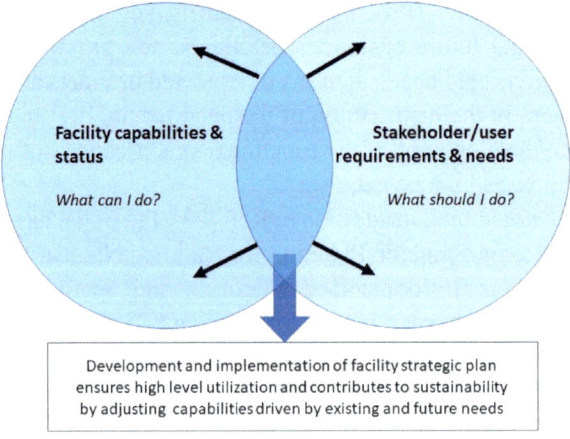

FIG. 3. Strategy for an IRF project to ensure high level utilization (derived from Ref. [23]).

facility, it may contribute to the justification of the need for the IRF and eventual specifications and design of the IRF under consideration. In the case of an existing facility, as an example, this could avoid the risk of loss of corporate knowledge as ageing staff retire. During the planning process, while the emphasis might be on utilization and efficient operation, the need for safety and plant improvements as well as lifetime extension by equipment refurbishment should not be overlooked. A brief overview of the process will show the steps for a new facility and for an existing facility. It will be found that they are logically linked together.

The strategic planning methodology for a new IRF could involve the following steps:

(a) Determine the potential stakeholders and their quantified needs in the utilization of the facility.
(b) Determine the required capabilities of the new facility on the basis of these needs.
(c) Perform an iterative analysis that examines (a) and (b) in the context of the environment, available/expected resources and constraints or issues under which the facility will operate, in order to accomplish the following:
    — Generate a vision of future goals and major objectives;
    — Decide on a few specific objectives and the detailed action plans required to achieve them;
    — Implement, review progress and revise the plan.

The strategic planning methodology for an existing IRF could involve the following steps:

(a) Examine and quantify present and potential capabilities of the facility.
(b) Determine the existing and potential stakeholders and quantify their needs.
(c) Perform an iterative analysis that examines (a) and (b) in the context of the environment, available/expected resources and constraints or issues under which the facility will operate, in order to accomplish the following:
    — Generate a vision of future goals and major objectives;
    — Decide on a few specific objectives and the detailed action plans required to achieve them;
    — Implement, review progress and revise the plan.

One can observe that the approach for new and existing IRFs is quite similar; only the order of steps (a) and (b) is reversed. The following list provides some typical examples of constraints or issues to be examined:

— Priorities of major stakeholders such as the government, NNIs or investors;
— Utilization (e.g. priority and scheduling of R&D time vs. provision of commercial products and services);
— Funding and financing, including return on revenue;
— Outreach and marketing;
— Alternative technologies and competitors;
— Regulatory matters;
— Human resource and knowledge management;
— Radioactive waste management.

During step (c) of strategic plan development, SWOT (strengths, weaknesses, opportunities, threats) analysis and probabilistic risk assessment (PRA) via a risk matrix are commonly used tools [22, 23, 37, 38]. A SWOT evaluation is an analysis methodology that can be used to aid decision making in the case of both existing and new IRFs. Clearly, this is not an isolated process, as one needs to assess what resources are available or could be available, and what realistic risks might prevent any proposed implementation. This information is then put into context and analysed within the framework of the environment and constraints or issues applicable to the facility. This is when the probabilistic risk assessment becomes useful for the evaluation of the identified risks and determines the likelihood of the events occurring and their consequences. The assessment will also help the identification of countermeasures to prevent the occurrence of risks.

The SWOT and probabilistic risk assessment analyses lead to specific conclusions and the ultimate definition of proposed strategies. These, in turn, lead to some major objectives for the proposed IRF, as illustrated in Fig. 4, becoming the central focus of future efforts of the IRF's PSP. Each major objective then usually cascades into one or more specific objective(s) with detailed action plans for achieving them. This approach is explained in detail in section 3.8 and annex II to Ref. [23] (and can also be understood from Ref. [35]).

3.1.3. Contents of a preliminary strategic plan document

The contents of the PSP for IRFs may differ for each project or organization and will need to be adjusted to specific facilities and their respective goals. On the other hand, the table of contents illustrated in Fig. 5 can serve as a reference template and may be applied to most IRFs.

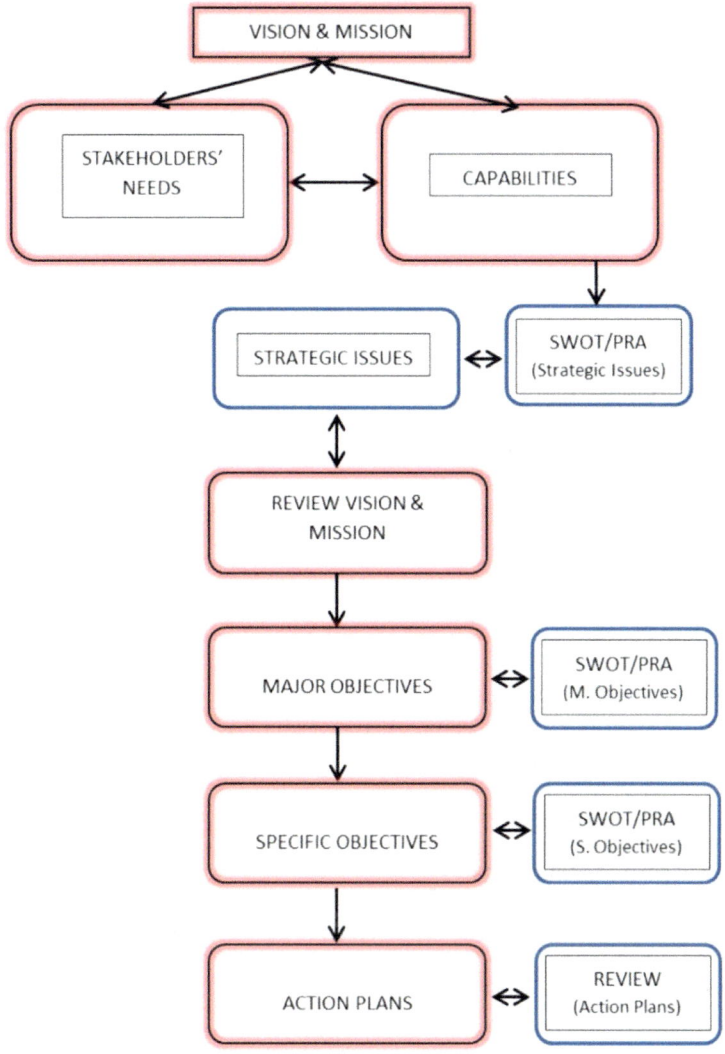

FIG. 4. Outline of the strategic planning methodology (reproduced from Ref. [23]).

CONTENTS   page

FIG. 5. An example of contents of a PSP document (reproduced from Ref. [23]).

3.2. ASSESSMENT OF INFRASTRUCTURE

3.2.1. Objectives

The scope of the infrastructure assessment includes both the 'hard' (facilities, equipment, etc.) and 'soft' (legal and regulatory, training, quality management, etc.) infrastructure items needed for an IRF as well as the evolution of these needs from the time an organization considers an IRF and its associated facilities, through all stages of planning, decision making and implementation.

The main purpose of an infrastructure assessment is to identify what infrastructure elements or issues (see Table 2 for details) might be already in place and which still need to be developed and implemented while the establishment of a new IRF is being considered.

3.2.2. Methodology for infrastructure assessment

Self-assessment of infrastructure is encouraged as the first step in any overall review of readiness to proceed to the next phase of the development of infrastructure for an IRF. Self-assessment is also an essential tool for continual improvement. Although a self-assessment will be performed by an organization planning an IRF and will include persons that are part of the multiple stakeholders involved in the project, consideration needs to be given to augmenting the PIT by involving consultants and independent experts from within the organization or from outside. The key requirement for any assessment is to assign people, internally or externally, who have a thorough understanding of infrastructure issues and have knowledge and experience in conducting assessment activities.

It is vital that an integrated assessment be carried out across all infrastructure issues (see Table 2 and Annexes I and II) because every one of them is essential, and there is significant linkage among them. Accordingly, there is a need to fully integrate the management of each infrastructure issue and associated human and financial resources. As mentioned in Sections 2.1 and 2.5, the assessment methodology for the issues that have a strong connection to the existing national framework can be simplified.

It is also necessary to understand the purpose of the assessment made at project milestone 1. The assessment is about ensuring the quality of information available and the effective investment of resources for informed decisions and the management of risks associated with the entire project. Even though an entity can do only limited work in phase 1, it might carry a much greater risk of making a decision that is not well-informed, or of phase 2 taking much longer than planned because the necessary infrastructure issues have not been properly identified and addressed, or of low utilization of the new facility because the involvement of utilization stakeholders was insufficient.

An entity considering building its very first IRF, or a new type of IRF which is yet to confirm its sound standing, is encouraged to make significant use of the available experts and good practices of other organizations. The use of partnership agreements with other local and foreign organizations will be beneficial for the success of the project. However, it is important to note that responsibility for readiness to proceed to the next phase lies solely with the organization embarking on the IRF project.

3.2.3. Evaluation steps

All infrastructure issues will be examined in the self-assessment to obtain a complete picture of the status and progress of the required infrastructure development. In principle, a self-assessment can be carried out at any time, but this publication assumes that the initial self-assessment will be carried out at the beginning of phase 1 and that identified gaps will be addressed and closed during phases 2 and 3.

A comprehensive assessment comprises four main steps:

(a) Defining the terms of reference for the assessment, including identification of the organizations to be involved and the team/individuals who will conduct the assessment;
(b) Evaluating the status of development of the IRF infrastructure against the conditions set, as listed in Annex I to this publication;
(c) Identifying areas/gaps which need further attention;
(d) Preparing a corresponding action plan to address these areas and fill the gaps.

It is beneficial that all these steps be undertaken to obtain comprehensive and accurate information about whether the organization has completed the work across all the infrastructure issues and to identify any infrastructure gaps and outstanding work (and resources) to be addressed.

3.2.4. Documenting the results and preparing an action plan

Following the self-assessment process, the entity has to prepare a self-evaluation report (SER). It is suggested that the SER contain, as a minimum, the following elements:

— Identification of the 'team of evaluators' by position/role in their respective organizations;
— Identification of the 'team of respondents';
— Description of the process used to conduct the assessment;
— Lists of the evidence reviewed, and further actions required;
— Summary and conclusions giving the state of achievement of each issue/condition;
— References to any relevant material used for conducting the assessment;
— Confidentiality requirements, if any.

A tabular format such as Fig. 6 is an example of collating and summarizing the results of the assessment in the SER carried out for each condition related to each infrastructure issue. To assess overall progress and to assign priorities, each condition is given a 'status'. Three options are suggested: (a) significant actions needed; (b) minor actions needed; (c) no actions are needed.

Upon completion of the SER, an action plan needs to be developed. The observations from the SER will be used by the entity to determine this action

1. Organizational position	
Conditions	**Basis for evaluation**
1.1. Long term commitment made and importance of safety and security recognized	**Summary of the condition to be demonstrated** A clear statement adopted by the organization/institution/company (and if relevant, by a governmental authority) on its intent to establish a new IRF and of its commitment to safety, security and sustainability, with evidence that its importance is embedded in the ongoing development programme or strategy.
	Examples of how the condition may be demonstrated — A clearly stated commitment by the organization (and governmental authority if relevant), including the prime responsibility for safety; — Evidence of clear responsibilities for each infrastructure issue, within the organization and other relevant authorities.
	Selected relevant IAEA publications IAEA Nuclear Energy Series No. NG-T-3.18, Feasibility Study Preparation for New Research Reactor Programmes, IAEA (2018) IAEA, Governmental, Legal and Regulatory Framework for Safety, IAEA General Safety Requirements No. GSR Part 1 (Rev.1), IAEA (2016) IAEA, Radiation Protection and Safety of Radioactive Sources: International Basic Standards, IAEA General Safety Requirements No. GSR Part 3, IAEA (2014)

FIG. 6. Example of a self-assessment form for a selected infrastructure issue (1) and one of the conditions (1.1).

plan. Each entity will decide the most appropriate way to prepare the action plan, but it needs to include the following:

— The issue being addressed;
— A clear statement of the actions to be taken, showing how it will address the identified shortfall;
— The definition of the objectives, according to the 'SMART' indicator (specific, measurable, achievable, relevant and timebound) approach, to be achieved under each action to fulfil the conditions of the issue;
— An agreed completion time for each action;
— The organization/function/post holder responsible for the completion of the actions;
— The required resources and budget to complete the actions.

It is important that the actions be 'owned' by the organization responsible for their completion and that said organization identifies the resources (staff and budget) needed to complete the actions within the agreed time frame. In some cases, direct involvement of the governmental authorities may be required, with or without allocation of additional resources. Further details on the preparation of action plans and examples thereof can be found in Refs. [23, 35].

3.3. COST–BENEFIT ANALYSIS AND BUDGETARY ESTIMATION

3.3.1. Objectives

The objective of the cost–benefit analysis is to ensure that the budgetary considerations provided in the FSR include an estimation of the overall lifetime costs and monetary income (or tangible impact for facilities that are not commercially oriented), and these need to be presented with appropriate margins for all the project phases.

3.3.2. Methodology

The cost–benefit analysis of an IRF is the comparison of its lifetime cost with its benefits (income or other tangible impacts). For a specific IRF, the tailored approach is recommended. To support and structure cost estimates, the costs and benefits associated with an IRF project need to be split into logically ordered categories. In addition to the major capital investment costs, other costs may include (but are not limited to) the following:

— Feasibility study: forming teams and ensuring associated training, preparing a PSP, performing self-assessment of required infrastructure, performing cost–benefit analysis, selecting preliminary candidate site(s), consulting services and subcontracting of external experts.
— Bidding process: preparing requests for proposals, quotation or tenders, evaluation, contracting and follow-up activities.
— Human resources, including human resources development to cover identified necessary competence needs and knowledge management costs: staff including administrative, technical and managerial personnel for the future operating organization and other authorities[13], for project management and implementation during the planning and development stages, and salaries and overhead costs for operational staff during IRF utilization.
— Siting or space allocation: costs associated with surveys, characterization and procurement, also with ground preparation and maintenance of the facility (e.g. provision of additional roadways, power, water, drainage, fence/boundaries). In some cases, environmental impact assessment also might be required.
— External technical support: various technical services will be required during the lifetime of the project and some of these may be outsourced, either in part or in their entirety.
— Legal counselling and other professional services: lawyers, economists, market analysts and other professional staff will likely be required on an ad hoc basis.
— Construction work, which often includes the largest cost categories: people, materials and equipment costs for the construction of site, buildings and other infrastructure and for the IRF, as well as associated facilities and support. These may also be required for subsequent expansion of the facilities.
— Commissioning: costs associated with developing a commissioning plan, functional testing and acceptance, verifying that staff training and operation and maintenance manuals comply with requirements, licensing fees (where applicable), insurance and others.
— Future operation and maintenance: ongoing costs for the purchase of equipment and consumables need to be estimated on a yearly basis, taking into consideration a reference utilization scheme.

---

[13] The regulatory body should be independent and therefore costs of its human resources development are not included in the cost–benefit analysis.

— Outreach and marketing (for service oriented IRFs, including competitor analysis on an ongoing basis).
— Safety assessments and licensing, compliance with regulatory requirements and authorizations.
— Radioactive waste management, including insurance for or prepaid disposal of disused radioactive sources in the event the operator goes bankrupt.
— Decommissioning, including site cleanup as required.[14]
— Security arrangements and assessments, including during use, storage and transportation of radioactive materials.

The possible financial incomes need to be evaluated on the basis of the planned utilization strategy set out in the PSP. The income may include the following:

— Revenues from the provision of analytical and consultancy services;
— Revenues from the product sales or services delivered;
— Committed subsidies/donations from the funding authorities;
— Fees for the use of the IRF and its associated facilities;
— Revenues from the provision of education, training and quality management.

In addition, considering the diversity in motivation, ownership arrangements and organizational interest in investing in an IRF, one may have to consider intangible externalities that would accrue over relatively long timescales, for example, education and health benefits to the general public. In that case, financial evaluations and assessments established only on the basis of easily quantified expenditures, income and depreciation of tangible assets may not be sufficient. Consequently, the feasibility of an IRF project cannot be best assessed purely on the basis of simple financial matrices and projections that do not take into account the larger benefits to academia, public health, society or trade exports. Rather, the broader potential social, technological and economic benefits need to be evaluated. However, these benefits may not be conducive to simple quantification. For these circumstances, the relevant national authority or PMT may authorize another analytical method to be applied.

Projections of revenue generation with a new commercially oriented IRF need to take into account the additional time and resources required to develop all the capabilities and, if relevant, the necessary licensing and approval processes (e.g. the drug master file for medical radioisotopes and associated

---

[14] In the case of IRF projects involving the use of radioactive sources of Category 1 or 2 (desirable for Category 3, too), the cost of end-of-life management of the sources (and operational strategy for the same) should be considered.

radiopharmaceuticals; the accreditation of quality management systems, etc.), which may take a long time, from several months to years after the IRF becomes operational. Necessary efforts and possible delays should not be underestimated. In addition to this, market penetration with a new product or service will invariably take time and also depends on developments with competitors, if any. As such, it may be misleading to assume that full income generation will start as soon as the IRF is commissioned. It is therefore better to stay conservative with expectations on income generation in the first few years following the commissioning of the facility and to develop an income generation plan that is graded over several years on the basis of operational experience. Information from similar facilities, national or regional, may help to get a balanced view on what may be achieved in reality.

It is also important to clarify that income generation requires resources such as personnel, consumables, spare parts and maintenance contracts that have to be covered by the revenues. An overly optimistic cost–benefit analysis may quickly be interpreted as a 'money generator', whereas the profit margin has to be kept low in view of actual penetration in already existing and possibly saturated markets.

It is important to recognize that, even for a basic financial analysis such as what would be expected in an FSR, a cost and revenue estimation exercise is complex. There are many alternative techniques and tools that could be employed to support the generation of the overall cost and benefit estimates required for the feasibility study. Evaluating these is beyond the scope of the current publication, but just to mention one example, the benefit–cost ratio methodology described in Refs. [36, 39] can be employed for this purpose.

UNIDO's Computer Model for Feasibility Analysis and Reporting (COMFAR) [40] includes cost–benefit and value-added methods of economic analysis and has been extensively employed to facilitate short term and long term analysis of financial and economic consequences for both industrial and non-industrial projects.

3.4. CONTENTS OF THE FEASIBILITY STUDY REPORT

The PSP, IAR and cost–benefit analysis should be integrated to prepare the FSR for the IRF in consideration. The contents of an FSR of a new IRF can be adopted from annex I to Ref. [22]. The same is reproduced here in Table 3 for the convenience of the reader.

TABLE 3. EXAMPLE OF TABLE OF CONTENTS OF FEASIBILITY STUDY REPORT

Title	Contents
Executive Summary	A short, informative and convincing summary of the FSR for the decision makers.
Chapter 1. Introduction	
1.1. Objectives	The objectives of a feasibility study and motivations of an IRF project.
1.2. Identification and quantification of stakeholder needs	Description of stakeholders and their needs in reflecting their main requirements using the PSP.
1.3. Opportunities	Description of how the IRF will contribute to the realization of vision and missions of the organization planning the project using the PSP.
1.4. Summary of the preliminary strategic plan	Summary description of the strategic considerations and main objectives as per the PSP.
1.5. Description of the IRF and its utilization	The functional requirements derived from the quantification of stakeholder needs and the description of the IRF as described in the PSP.
1.6. Summary of commitment	The commitment of the promoting institution for the IRF project. If the letters of intention for use from stakeholders are available, it is recommended to include them as well.
Chapter 2. Methodology for Feasibility Study	
2.1. Scope of analysis, assumptions and constraints	A description of the activities resulting from the feasibility study to convince the decision makers to support the project.
2.2. Employed methods	A summary of the major activities for preparing the PSP and conducting an infrastructure assessment and cost–benefit analysis.

TABLE 3. EXAMPLE OF TABLE OF CONTENTS OF FEASIBILITY STUDY REPORT (cont.)

Title		Contents
2.3.	Detailed demand studies	A description of how the demand study was performed to quantify the stakeholder needs using the PSP.
2.4.	Technical options studied	A description of how different IRFs, including alternative technologies, were considered, assessed and selected as one of the possible options to fulfil quantified needs of stakeholders/users.
2.5.	Assessment of national and organizational infrastructure issues and conditions for phase 1	A description of how the infrastructure assessment was conducted and the status of actions to fill the gaps identified in the pre-project phase using the IAR.
2.6.	Understanding of national and organizational infrastructure gaps for phases 2 and 3	The recognition of what should be performed for the preparedness of infrastructure in Phases 2 and 3 using the IAR.
2.7.	Uncertainties, missing information and assumptions	Identification of factors not fully known and the corresponding assumptions made, which can be taken from the PSP.
2.8.	Alternatives for project management	A description of alternatives for project management such as bidding, design, construction and commissioning of the IRF.
2.9.	Facility installation and decommissioning options	The options to be considered for operation of the IRF and its eventual decommissioning.
2.10.	Cost–benefit analysis model	The cost–benefit model used and analysis conducted.
2.11.	Risk analysis	The risk analysis conducted (including the input provided in Section 2.7), which can be taken from the PSP.

TABLE 3. EXAMPLE OF TABLE OF CONTENTS OF FEASIBILITY STUDY REPORT (cont.)

Title	Contents
Chapter 3. Feasibility Study Results	
3.1. Results of the service and product demand studies	The outputs from the actions described in Section 2.3.
3.2. Technical solutions	A description of the facility similar to that in Section 1.5 for fulfilling the demands shown in Section 3.1.
3.3. Human resource development and project management	The results from the actions to complete the relevant infrastructure issue and the choice of project management method deduced from the alternatives in Section 2.8 will be described.
3.4. Analysis of gaps from infrastructure assessment	The status of infrastructure issues which are the outcome of actions for phase 1. In addition, it is recommended that the major action plans for phases 2 and 3 be described. These are to be taken from the IAR.
3.5. Financial considerations	The financial requirements and multi-year budget plan developed using the results from the activities for Section 2.10.
3.6. Project schedule proposal	The project schedule/Gantt chart to design, construct and commission the IRF.
3.7. Business model and its implementation	The outreach and marketing schemes to realize the service and product described in Section 3.1 or PSP.
3.8. Risk management	The result of risk analysis performed by following Section 2.11 and the measures to prevent the occurrence of risks, which can be taken from PSP.
Chapter 4. Summary	The summary of the main feasibility study activities and results, including justification of the IRF.

4. CONCLUDING REMARKS

This publication provides a sound and systematic approach for consideration, if not adoption, when planning, establishing and utilizing an IRF. The process starts from an objective analysis of the justification of the need for the IRF, moves to a detailed, comprehensive assessment of infrastructure requirements and means of addressing them, continues with a cost–benefit analysis and then finishes with the development of a PSP for the safe, secure and effective utilization of the IRF. The accrual of benefits from the IRF ranges from aiding education and research, supporting healthcare and patient management, industrial process efficiency and quality to being an enabler of exports, food preservation, advanced materials and their characterization and cultural heritage, among others. The resultant FSR on the lines of the recommended structure (see the template given in Table 3) will facilitate and enable well-informed decision making regarding the proposed IRF establishment and the subsequent steps to be followed.

A three-phase project milestones approach (Table 1) was developed, adapted with suitable modifications and tailoring from the Milestones Approach developed by the IAEA for nuclear power plants and later adapted to research reactors. However, interest in IRFs is high in almost all countries, and also goes well beyond NNIs and government entities. This is a significant difference compared to the case of nuclear power plants and research reactors. Moreover, other regulatory entities, beyond those responsible for nuclear and radiation safety and security, may be involved in the case of certain IRFs. Accordingly, the tables in Annex I on the assessment of infrastructure conditions reflect all of the above aspects. In addition, IRFs, by their nature, type, degree of complexity and associated potential risks, taken along with their specific features, have widely variable requirements and issues. Therefore, further grading and tailoring will invariably be necessary in applying the guidance given in this publication. This will be particularly important when applying the conditions enumerated in the tables of Annex I on the assessment of infrastructure and confirming their degree of fulfilment.

It is quite likely that the need for IRFs is strongly felt even in countries that have no or only limited familiarity or expertise in nuclear and radiation applications and, accordingly, may not necessarily have a formal regulatory framework and other necessary infrastructure in place. This publication would be of high value in such cases for both the government and the national entities planning the IRF based pursuits. On the other hand, in many countries, private entities, including commercial ones, are interested in setting up IRFs outside of NNIs. This is a distinct feature of many IRFs (as compared, for example, to research reactors), and this scope is covered under the 16th issue (fostering expansion of ownership) in the assessment of infrastructure conditions (see Table 2 and Annex I).

Furthermore, it is recognized that, in some countries, certain decisions regarding IRFs, especially those related to healthcare and research (academia), may follow (or could have followed, for existing IRFs) a direct *top-down* approach, namely 'an executive decision has been made, and implementation is already underway'. Patient care needs and aspirations of leaders of academic research, respectively, could have (understandably) driven and accelerated the national or organizational decision making in the above cases.

Above all, there are hundreds (if not thousands) of IRFs already in operation in a large number of countries across the world (and more being set up), including some which may not be in regular use for one reason or another. In all these cases, the current publication will be useful for one or more of the following purposes: retrofitting by objective analysis; troubleshooting and enhanced utilization; upgrading or refurbishing; addressing viability and sustainability of operations; (needs based) midcourse refinements; strengthening effective and sustainable management; expansion of an existing IRF; propagation of an IRF by facilitating adoption; and enhancing safety and security. The publication is thus expected to be of high interest and utility to a diverse spectrum of entities and stakeholders in IAEA Member States.

REFERENCES

[1]   INTERNATIONAL ATOMIC ENERGY AGENCY, Nuclear Technology and Applications, IAEA, Vienna (2021),
https://www.iaea.org/topics/nuclear technology and applications

[2]   INTERNATIONAL ATOMIC ENERGY AGENCY, Sustainable Development Goals (SDGs), IAEA, Vienna (2021),
https://www.iaea.org/about/overview/sustainable development-goals.

[3]   INTERNATIONAL ATOMIC ENERGY AGENCY, Sustainability and Self Reliance of National Nuclear Institutions, Proc. of a Workshop Held in Daejeon, Republic of Korea, 9–12 April 2019, IAEA TECDOC 1943, IAEA, Vienna (2021).

[4]   INTERNATIONAL ATOMIC ENERGY AGENCY, Milestones in the Development of a National Infrastructure for Nuclear Power, IAEA Nuclear Energy Series No. NG G 3.1 (Rev.1), IAEA, Vienna (2015).

[5]   INTERNATIONAL ATOMIC ENERGY AGENCY, Specific Considerations and Milestones for a Research Reactor Project, IAEA Nuclear Energy Series No. NP T 5.1, IAEA, Vienna (2012).

[6]   VERZUH, E., The Fast Forward MBA in Project Management, 3rd edn, John Wiley & Sons, Hoboken, NJ (2008) p. 149.

[7]   MARIO LOPEZ, Milestones Scheduling Approach, Project Managers, International Institute of Executive Careers, Buffalo, USA (2020),
        https://projectmanagers.org/management/time/milestone-scheduling-approach/

[8]   INTERNATIONAL ATOMIC ENERGY AGENCY, Decommissioning of Facilities, IAEA Safety Standards Series No. GSR Part 6, IAEA, Vienna (2014).

[9]   INTERNATIONAL ATOMIC ENERGY AGENCY, Categorization of Radioactive Sources, IAEA Safety Standards Series No. RS-G-1.9, IAEA, Vienna (2005).

[10]   NON-ISOTOPIC ALTERNATIVE TECHNOLOGIES WORKING GROUP, Non-Isotopic Alternative Technologies White Paper (2019), Cybersecurity and Infrastructure Security Agency, US Department of Homeland Security, Washington, DC
        https://armscontrolcenter.org/wp-content/uploads/2019/10/Non-Radioisotopic-Alternative-Technologies-White-Paper1491.pdf

[11]   WORLD INSTITUTE FOR NUCLEAR SECURITY, Considerations for the Adoption of Alternative Technologies to Replace High Activity Radioactive Sources, WINS Special Reports Series (2021).

[12]   NATIONAL ACADEMIES OF SCIENCES, Report on Radioactive Sources: Applications and Alternative Technologies, The National Academies Press, Washington, DC (2021).

[13]   INTERNATIONAL ATOMIC ENERGY AGENCY, Governmental, Legal and Regulatory Framework for Safety, IAEA Safety Standards Series No. GSR Part 1 (Rev. 1), IAEA Vienna (2016).

[14]   INTERNATIONAL ATOMIC ENERGY AGENCY, Radiation Protection and Safety of Radioactive Sources: International Basic Standards, IAEA Safety Standards Series No. GSR Part 3, IAEA, Vienna (2014).

[15]   INTERNATIONAL ATOMIC ENERGY AGENCY, Nuclear Security Recommendations on Radioactive Material and Associated Facilities, IAEA Nuclear Security Series No. 14, IAEA, Vienna (2011).

[16]   INTERNATIONAL ATOMIC ENERGY AGENCY, Nuclear Security Recommendations on Nuclear and Other Radioactive Material out of Regulatory Control, IAEA Nuclear Security Series No. 15, IAEA, Vienna (2011).

[17]   INTERNATIONAL ATOMIC ENERGY AGENCY, Security of Radioactive Material in Transport, IAEA Nuclear Security Series No. 9 G (Rev. 1), IAEA, Vienna (2020).

[18]   INTERNATIONAL ATOMIC ENERGY AGENCY, Security of Radioactive Material in Use and Storage and of Associated Facilities, IAEA Nuclear Security Series No. 11 G (Rev. 1), IAEA, Vienna (2019).

[19]   INTERNATIONAL ATOMIC ENERGY AGENCY, IAEA Nuclear Safety and Security Glossary: 2022 (interim) Edition, IAEA, Vienna (2022).

[20]   INTERNATIONAL ATOMIC ENERGY AGENCY, Code of Conduct on the Safety and Security of Radioactive Sources, IAEA, Vienna (2004).

[21] INTERNATIONAL ATOMIC ENERGY AGENCY, Measures to Strengthen International Cooperation in Nuclear, Radiation and Transport Safety and Waste Management, GC (48)/RES/10, IAEA, Vienna (2007).

[22] INTERNATIONAL ATOMIC ENERGY AGENCY, Feasibility Study Preparation for New Research Reactor Programmes, IAEA Nuclear Energy Series No. NG-T-3.18, IAEA, Vienna (2018).

[23] INTERNATIONAL ATOMIC ENERGY AGENCY, Strategic Planning for Research Reactors, IAEA Nuclear Energy Series No. NG-T-3.16, IAEA, Vienna (2017).

[24] INTERNATIONAL ATOMIC ENERGY AGENCY, Specific Considerations in the Assessment of the Status of the National Nuclear Infrastructure for a New Research Reactor Programme: Reference document for the INIR RR missions, IAEA Nuclear Energy Series NR-T-5.9, IAEA, Vienna (2021).

[25] INTERNATIONAL ATOMIC ENERGY AGENCY, Technical Requirements in the Bidding Process for a New Research Reactor, IAEA Nuclear Energy Series No. NP-T-5.6, IAEA, Vienna (2014).

[26] PROJECT MANAGEMENT INSTITUTE, A Guide to the Project Management Body of Knowledge (PMBOK guide), 5th edn, Newtown Square, PA (2013).

[27] AXELOS, Managing Successful Projects with PRINCE2, 6th edn, The Stationery Office, Norwich (2017).

[28] PROJECT MANAGEMENT INSTITUTE, The Standard for Program Management, 4th edn, Newtown Square, PA (2017).

[29] INTERNATIONAL ATOMIC ENERGY AGENCY, Responsibilities and Capabilities of a Nuclear Energy Programme Implementing Organization, IAEA Nuclear Energy Series NG-T-3.6 (Rev. 1), IAEA, Vienna (2019).

[30] Review Missions and Advisory Services, IAEA, Vienna, https://www.iaea.org/services/review-missions

[31] Nucleus Information Resources, IAEA, Vienna, https://www.iaea.org/resources/nucleus-information-resources

[32] INTERNATIONAL ATOMIC ENERGY AGENCY, Use of a Graded Approach in the Application of the Management System Requirements for Facilities and Activities, IAEA TECDOC No. 1740, IAEA, Vienna (2014).

[33] INTERNATIONAL ATOMIC ENERGY AGENCY, Fundamental Safety Principles, IAEA Safety Standards Series No. SF-1, IAEA, Vienna (2006).

[34] INTERNATIONAL ATOMIC ENERGY AGENCY, Safety Glossary, Terminology Used in Nuclear Safety and Radiation Protection 2022 (interim) edn, IAEA, Vienna (2022).

[35] INTERNATIONAL ATOMIC ENERGY AGENCY, Strategic Planning for National Nuclear Institutions, E-learning course available through CLP4NET, IAEA, Vienna (2019), https://elearning.iaea.org/m2/course/index.php?categoryid=124

[36] INTERNATIONAL ATOMIC ENERGY AGENCY, Análisis Probabilista de Seguridad de Tratamientos de Radioterapia con Acelerador Lineal, IAEA TECDOC No. 1670, IAEA, Vienna (2012).

[37] INTERNATIONAL ATOMIC ENERGY AGENCY, Application of the Risk Matrix Method to Radiotherapy, IAEA TECDOC No. 1685, IAEA, Vienna (2016).

[38] KOREA DEVELOPMENT INSTITUTE, General Guidelines for Preliminary Feasibility Studies, 5th edn, Public and Private Infrastructure Investment Management Centre, KDI, Seoul (2008),
https://www.kdi.re.kr/kdi_eng/publications/publication_view.jsp?pub_no=13070

[39] ALBY, T., Science and Project Management, (2020),
https://project-management-knowledge.com/

[40] UNITED NATIONS INDUSTRIAL DEVELOPMENT ORGANIZATION, Computer Model for Feasibility Analysis and Reporting (COMFAR), UNIDO, Vienna, (2020),
https://www.unido.org/resources/publications/publications-type/comfar-software

Annex I

BASIS FOR INFRASTRUCTURE ASSESSMENT

The minimum conditions for each infrastructure issue to be achieved before the IRF enters the operational phase are described in the following series of tables, which include examples of evidence to demonstrate the fulfilment of these conditions. Attention is drawn to the discussion on the graded approach in Section 2 of this publication.

The types of IRF targeted by this guide are numerous as well as diverse (see Section 1.4 for details) but they do not fall under the category of nuclear facilities. Therefore, it is recommended to interpret the contents of the tables, applying due care and keeping in mind the specific nature of the IRF under consideration, and also applying a graded approach and tailoring as appropriate to the nature and complexity of the IRF.

It is necessary for an appropriate national legal and regulatory framework in line with international standards and recommendations for radiation safety and the security of radiation sources to be in place well in advance, prior to the consideration of an IRF project. As mentioned in Section 2.1, several infrastructure issues have a strong connection to the national framework. These issues are nuclear safety, legal framework, safeguards, regulatory framework, environmental protection (particularly in those aspects that concern the national framework) and nuclear security. A condition to develop a successful IRF project is that no major gaps in the infrastructure development for those issues exist, as the national framework is, in most cases, not expected to require significant changes for the purpose of establishing an IRF. For those issues, the conditions in this Annex can usually be met in a simplified way, by referring to the results of IAEA review missions and advisory services or by referring to existing IAEA databases and other electronic information resources. A list of these services and resources is given in Annex II.

The tables below refer to 'evidence' and 'plans'. Evidence can include laws, decrees, regulations, formalized agreements, contracts, reports, meeting notes, correspondence, presentations, conferences attended with meeting reports, discussions held with minutes, personnel curricula, organization descriptions, job descriptions, etc. Plans need to have clear actions with associated time frames, resources required, and evidence that they are available as well as predefined measurable performance indicators. In all cases, documents need to be vetted by a person/organization with the appropriate authority.

There are many ways to establish and manage an IRF project (e.g. own country lead contractor, turnkey, multipackage contract, private-public enterprise). This publication does not seek to prescribe a particular approach and therefore can be applied as a general methodology, since the requirements do not change. What may change in some cases is how the conditions are fulfilled, and this is recognized within the detail of the proposed basis.

On completion of the comprehensive assessment, what is clearly needed is strong evidence of a holistic approach to information gathering, analysis, resource development and decision making. This view will be obtained by addressing each of the infrastructure issues and then integrating them into the overall IAR.

1. Organizational position [a]	
Conditions	**Basis for evaluation**
1.1. Long term commitment made and importance of safety and security recognized	**Summary of the condition to be demonstrated** A clear statement adopted by the organization/institution/company (and if relevant, by a governmental authority) on its intent to establish a new IRF and its commitment to safety, security and sustainability, with evidence that their importance is embedded in the ongoing development programme or strategy.
	Examples of how the condition may be demonstrated — A clearly stated commitment by the organization (and governmental authority if relevant), including the prime responsibility for safety; — Evidence of clear responsibilities for each infrastructure issue within the organization and other relevant authorities.
	Selected relevant IAEA publications Feasibility Study Preparation for New Research Reactor Programmes, Nuclear Energy Series No. NG-T-3.18, IAEA, Vienna (2018) Governmental, Legal and Regulatory Framework for Safety, IAEA General Safety Requirements No. GSR Part 1 (Rev.1), IAEA, Vienna (2016) Radiation Protection and Safety of Radioactive Sources: International Basic Standards, IAEA General Safety Requirements No. GSR Part 3, IAEA, Vienna (2014)
1.2. Project management and implementation teams established	**Summary of the condition to be demonstrated** Project management (PMT) and project implementation (PIT) teams — Have clear terms of reference that call for a comprehensive review of all the issues relevant to making a decision to proceed with a new IRF project; — Are recognized by all relevant authorities as having that role; — Have appropriate competencies and human and financial resources; — Involve all relevant stakeholders, including the new IRF users, the regulatory body for security of radioactive materials and radiation safety, other relevant government agencies, legislative representatives and other decision makers.

[a] A condition to develop a successful IRF project is that no major gaps in the infrastructure development for this issue exist, as the national framework is, in most cases, not expected to require significant changes for the purpose of establishing an IRF. This issue is usually checked by referring to the results of IAEA review missions and advisory services or by referring to existing IAEA database and other electronic information resources. A list of these services and resources is given in Annex II.

1. Organizational position (cont.)	
Conditions	**Basis for evaluation**
	Examples of how the condition may be demonstrated — The charter establishing the PMT and PIT, and to whom they report; — Evidence that the roles and responsibilities of the PMT and PIT are known by all their members and by other relevant authorities; — A document defining objectives and timescales and an adequate scope of investigations; — A clear description of how the PMT and PIT operate in terms of funding, planning, reporting, scope of studies and use of consultants; — Evidence that the PMT and PIT have adequate skills to address all issues either directly or through commissioning specialist studies; — Evidence of relevant interactions between the heads of PMT and PIT, and relevant authorities.
	Selected relevant IAEA publications Feasibility Study Preparation for New Research Reactor Programmes, Nuclear Energy Series No. NG-T-3.18, IAEA, Vienna (2018)
1.3. Comprehensive feasibility study performed, documented and the necessary commitments understood	**Summary of the condition to be demonstrated** A comprehensive FSR, defining and justifying a new IRF; this report will incorporate and update the assessment of national/organizational infrastructure, PSP and cost–benefit analysis and will integrate these with the analysis of the obligations, commitments and resources required.
	Examples of how the condition may be demonstrated — List of the studies that are feeding into the comprehensive FSR; — Contents list and executive summary of the report(s); — Evidence that PMT and relevant authorities reviewed of the report(s).
	Selected relevant IAEA publications Feasibility Study Preparation for New Research Reactor Programmes, Nuclear Energy Series No. NG-T-3.18, IAEA, Vienna (2018)

2. Nuclear and radiation safety	
Conditions	**Basis for evaluation**
2.1. Key requirements of safety and international guidance and practices understood	**Summary of the condition to be demonstrated** The key requirements for safety, in line with the IAEA safety standards and relevant to IRF, are understood by the PMT and PIT and other relevant stakeholders, and their implications are recognized. Consultation with the Regulatory Body for Safety is demonstrated. The need for international cooperation and open exchange of information related to radiation safety as an essential element is recognized.
	Examples of how the condition may be demonstrated — Evidence that the PMT and PIT have an understanding of and commitment to safety; — Evidence that the responsibility for safety is recognized, for example, in consideration of leadership, expertise and funding; — Evidence that the need to develop adequate capability and skills in safety is recognized, including emergency preparedness and response (EPR); — Evidence of familiarity with the national legislative and regulatory system for safety; — Evidence that the prime responsibility for safety is assigned to the organization responsible for IRF; — Evidence of familiarity with IAEA safety standards, international and other States' practices, and recognition of the need for, and commitment to, the development of national safety standards; — Implementation of a relevant national technical cooperation project with the IAEA and evidence of government financial support; — A comprehensive safety assessment has been performed, including the validation process; — Routine verification of safety; — External audits and regulatory control are implemented.
	Selected relevant IAEA publications Fundamental Safety Principles, Safety Standards Series No. SF-1, IAEA, Vienna (2006) Governmental, Legal and Regulatory Framework for Safety, IAEA General Safety Requirements No. GSR Part 1 (Rev.1), IAEA, Vienna (2016) Safety Assessment for Facilities and Activities, Safety Standards Series No. GSR Part 4 (Rev. 1), IAEA, Vienna (2016) Radiation Protection and Safety of Radioactive Sources: International Basic Standards, General Safety Requirements No. GSR Part 3, IAEA, Vienna (2014) Establishing the Infrastructure for Radiation Safety, Safety Standards Series No. SSG-44, IAEA, Vienna (2018) Radiation Safety of Gamma, Electron and X Ray Irradiation Facilities, Safety Standards Series No. SSG-8, IAEA, Vienna (2010)

2. Nuclear and radiation safety (cont.)	
Conditions	**Basis for evaluation**
2.2. Provisions of the Code of Conduct on the Safety and Security of Radioactive Sources and its Supplementary Guidance are understood and considered	**Summary of the condition to be demonstrated** The provisions of the Code of Conduct on the Safety and Security of Radioactive Sources are understood and considered, including through the incorporation of IAEA safety standards and nuclear security guidance in the national safety and security regulations (particularly if the country has made a political commitment to the IAEA to do so). See Issue 5.
	Example of how the condition may be demonstrated Evidence that the provisions of the Code of Conduct on the Safety and Security of Radioactive Sources have been integrated into the project since the earliest stages (e.g. by demonstrating the use of the IAEA safety standards and nuclear security guidance or attending or participating in the related IAEA meetings/activities).
	Selected relevant IAEA publications Code of Conduct on the Safety and Security of Radioactive Sources, IAEA/CODEOC/2004, IAEA, Vienna (2004) Guidance on the Import and Export of Radioactive Sources, IAEA/CODEOC/IMO-EXP/2012, IAEA, Vienna (2012) Guidance on the Management of Disused Radioactive Sources, IAEA/CODEOC/MGT-DRS/2018, IAEA, Vienna (2018)

3. Management	
Conditions	**Basis for evaluation**
3.1. Need for appropriate management systems recognized	**Summary of the condition to be demonstrated** The organization/institution/company has a commitment to management systems that will ensure success and promote a safety and security culture, sustainability and the peaceful use of IRFs. There are plans to ensure that the knowledge gained by the PMT and PIT, including information from the supplier, is transferred to the relevant regulatory authority(ies) and the future owner/operator of the new IRF, and that systems are in place for the preservation of critical knowledge throughout the entire life cycle.
	Examples of how the condition may be demonstrated — Plans to ensure the appointment of managers with the appropriate training and experience to plan, procure, construct and operate a new IRF as well as to ensure the leadership and management of nuclear safety and security of radioactive material; — Evidence that the importance of safety and security culture in relevant organizations/authorities to be established is recognized; — Evidence that the importance of ensuring the peaceful use of ionizing radiation technology is recognized; — Evidence of a clear understanding of management system requirements; — A plan to implement management systems in future key organizations/authorities is consistent with the appropriate standards and guidance.
	Selected relevant IAEA publications Leadership and Management for Safety, Safety Standards Series No. GSR Part 2, IAEA, Vienna (2016) Governmental, Legal and Regulatory Framework for Safety, General Safety Requirements No. GSR Part 1 (Rev.1), IAEA, Vienna (2016) Application of Plant Information Models to Manage Design Knowledge through the Nuclear Power Plant Life Cycle, IAEA-TECDOC-1919, IAEA, Vienna (2020) Knowledge Loss Risk Management in Nuclear Organizations, Nuclear Energy Series No. NG-T-6.11, IAEA, Vienna (2017)

4. Funding and financing	
Conditions	**Basis for evaluation**
4.1. Strategies for funding and financing established	**Summary of the condition to be demonstrated** The organization/institution/company has defined mechanisms for funding a range of key activities that are specific to an IRF but may not be the responsibility of the owner/operator. The activities might include: — The government's stakeholder involvement in the project; — Siting and environmental protection activities; — Emergency preparedness and response, if relevant; — Education, training and research; — Any required improvements to the specific infrastructure issues, if such improvements are the government's responsibility; — Management, storage and disposal of radioactive waste, if relevant; — Decommissioning of the IRF after its service life. Potential options have been identified with financial and risk management strategies, which together: — Identify the role of the organization and relevant governmental authorities in financing a new IRF project; — Ensure the long term viability of the owner/operator to fulfil all their responsibilities.
	Examples of how the condition may be demonstrated — Clear statements of how the above areas will be funded, based on a consideration of options. — Evidence that the scale of the costs of each of these activities has been recognized. — A review of financing options and risk management strategies, considering the possible long term economics and risks associated with the IRF. This should include the extent of required funding, equity partners, possible borrowing, etc.
	Selected relevant IAEA publications Feasibility Study Preparation for New Research Reactor Programmes (2018), Nuclear Energy Series No. NG-T-3.18, IAEA, Vienna
4.2. Cost–benefit analysis completed	**Summary of the condition to be demonstrated** Cost estimates associated with the establishment, operation and maintenance, and decommissioning of new IRF are performed. Estimation of expected revenue or other value generation by providing products and services using IRF.
	Examples of how the condition may be demonstrated — Report on cost–benefit analysis completed; — Evidence that the PMT and PIT endorse the cost–benefit analysis results as meeting objectives and expectations of a new IRF.
	Selected relevant IAEA publications Feasibility Study Preparation for New Research Reactor Programmes, Nuclear Energy Series No. NG-T-3.18, IAEA, Vienna (2018)

5. Legal framework [a]	
Conditions	**Basis for evaluation**
5.1. Adherence or commitment to all relevant international legal instruments	**Summary of the condition to be demonstrated** The following instruments have been adhered to and/or committed to: — Convention on Early Notification of a Nuclear Accident (INFCIRC/335); — Convention on Assistance in the Case of a Nuclear Accident or Radiological Emergency (INFCIRC/336); — Joint Convention on the Safety of Spent Fuel Management and on the Safety of Radioactive Waste Management (INFCIRC/546); — Convention on the Physical Protection of Nuclear Material (INFCIRC/274/Rev.1) and Amendment thereto (INFCIRC/274/Rev.1/Mod.1); — Comprehensive safeguards agreement — based on The Structure and Content of Agreements Between the Agency and States Required in Connection with the Treaty on the Non-Proliferation of Nuclear Weapons (INFCIRC/153 (Corrected)); — Additional protocol — following the provisions of Model Protocol Additional to the Agreement(s) Between States(s) and the International Atomic Energy Agency for the Application of Safeguards (INFCIRC/540 (Corrected)); — Revised Supplementary Agreement Concerning the Provision of Technical Assistance by the IAEA; — The Code of Conduct on the Safety and Security of Radioactive Sources (IAEA/CODEOC/2004) and its Supplementary Guidance on the Import and Export of Radioactive Sources (IAEA/CODEOC/IMO-EXP/2012) and Guidance on the Management of Disused Radioactive Sources (IAEA/CODEOC/MGT-DRS/2018).
	Example of how the condition may be demonstrated Evidence of adherence and/or commitments to the relevant international legal instruments.
	Selected relevant IAEA publications IAEA Handbook on Nuclear Law, IAEA, Vienna (2003) IAEA Handbook on Nuclear Law: Implementing Legislation, IAEA, Vienna (2010)

---

[a] A condition to develop a successful IRF project is that no major gaps in the infrastructure development for this issue exist, as the national framework is, in most cases, not expected to require significant changes for the purpose of establishing an IRF. This issue is usually checked by referring to the results of IAEA review missions and advisory services or by referring to existing IAEA databases and other electronic information resources. A list of these services and resources is given in Annex II.

5. Legal framework (cont.)	
Conditions	**Basis for evaluation**
5.2. Existence of comprehensive national nuclear law	**Summary of the condition to be demonstrated** A comprehensive national nuclear law is in place. The law needs to: — Establish an independent nuclear regulatory body with adequate human and financial resources, and a clear and comprehensive set of functions; — Identify responsibilities for safety and security; — Formulate safety principles and rules (prime responsibility for safety and security, radiation protection, radioactive waste, decommissioning, emergency preparedness and response and the transport of radioactive material); — Formulate nuclear security principles; — Give appropriate legal authority to, and define the responsibilities of, the regulatory body and all competent authorities establishing a regulatory control system (authorization, inspection and enforcement, review and assessment, development of regulations and guides, and public information); — Implement IAEA safeguards (in particular for IRFs that involve the use or production of nuclear material); — Implement import and export control measures for nuclear, radioactive material and items.
	Example of how the condition may be demonstrated Evidence that a comprehensive nuclear law is enacted and in force.
	Selected relevant IAEA publications IAEA Handbook on Nuclear Law, IAEA, Vienna (2003) IAEA Handbook on Nuclear Law: Implementing Legislation, IAEA, Vienna (2010) Governmental, Legal and Regulatory Framework for Safety, IAEA General Safety Requirements No. GSR Part 1 (Rev.1), IAEA, Vienna (2016)
5.3. All other legislation affecting the new IRF and related activities reviewed	**Summary of the condition to be demonstrated** Legislation has been reviewed and amended as necessary to cover: — Environmental protection; — EPR; — Occupational health and safety of workers; — Protection of intellectual property; — Local land use controls; — Roles of national and local governments; — Stakeholders and public involvement; — National health, food, agriculture laws and regulations; — International trade and customs; — Financial guarantees and any other required financial legislation; — Relevant R&D.
	Example of how the condition may be demonstrated Presentation of a review identifying relevant laws and evidence that the necessary laws have been enacted or that there is a clear plan to enact them at the appropriate time.

6. Safeguards [a]	
Conditions	**Basis for evaluation**
6.1. If needed, terms of international safeguards agreement revised, and plans made to implement changes	**Summary of the condition to be demonstrated** Most IRFs do not involve the use of nuclear material. This condition applies to those that do. — If the Member State currently has concluded a small quantities protocol to its comprehensive safeguards agreement, assessment needs to be done if second condition applies; — The Member State has a comprehensive safeguards agreement with associated subsidiary arrangements in force with the IAEA; — If relevant to a new IRF, the Member State has a plan describing how the existing State Systems of Accounting for and Control of Nuclear Material will be strengthened or adjusted to deal with the increase of activities and resources, as well as the need for enhancement of capabilities.
	Examples of how the condition may be demonstrated — Plans for rescinding the small quantities protocol and/or for ratification of the additional protocol, including the actions that need to be taken, clear assignment of responsibilities and understanding of the resources and the required timescales; — Evidence that the need for outreach activities is recognized to ensure that all existing and future entities having to report to the State authority for safeguards are aware of their roles and obligations.
	Selected relevant IAEA publications Safeguards Implementation Guide for States with Small Quantities Protocols, Services Series No. 22, IAEA, Vienna (2013) Nuclear Material Accounting, Services Series No. 15, IAEA, Vienna (2008) Guidance for States Implementing Comprehensive Safeguards Agreements and Additional Protocols, Services Series No. 21, IAEA, Vienna (2016) Safeguards Implementation Practices Guide on Establishing and Maintaining State Safeguards Infrastructure, Services Series No. 31, IAEA, Vienna (2018) Safeguards Implementation Practices Guide on Provision of Information to the IAEA, Services Series No. 33, Vienna (2016)

---

[a] A condition to develop a successful IRF project is that no major gaps in the infrastructure development for this issue exist, as the national framework is, in most cases, not expected to require significant changes for the purpose of establishing an IRF. This issue is usually checked by referring to the results of IAEA review missions and advisory services or by referring to existing IAEA databases and other electronic information resources. A list of these services and resources is given in Annex II.

7. Regulatory framework [a]	
Conditions	**Basis for evaluation**
7.1. An adequate regulatory framework is in place	**Summary of the condition to be demonstrated** The regulatory framework for radiation safety and nuclear security of radioactive material is in place, it matches the overall plan for the new IRF, and includes: — An effectively independent competent regulatory body with clear authority, adequate human and financial resources, and strong government support; — Assignment of core safety and security regulatory functions for developing regulations, review and assessment, authorization, inspection, enforcement and public information; — Authority and resources to obtain technical support as needed; — A clear definition of the relationship of the regulatory body to other organizations (e.g. technical support organizations and environmental agency); — Clearly defined responsibilities of licensees; — Authority to implement international obligations, if applicable; — Authority to engage in international cooperation, if applicable; — Provisions to protect proprietary, confidential and sensitive information; — Provisions for stakeholder and IRF users' involvement and communication with the public. There are agreed terms of reference for each regulator and a clear definition of roles of, and interfaces with, other regulators. There is recognition of the need for integrating existing nuclear security and radiation safety regulations with new regulations for an IRF and development of new guides. Plans to develop competence are addressed under infrastructure issue No. 10, human resource development.

---

[a] A condition to develop a successful IRF project is that no major gaps in the infrastructure development for this issue exist, as the national framework is, in most cases, not expected to require significant changes for the purpose of establishing an IRF. This issue is usually checked by referring to the results of IAEA review missions and advisory services or by referring to existing IAEA databases and other electronic information resources. A list of these services and resources is given in Annex II.

7. Regulatory framework (cont.)	
Conditions	**Basis for evaluation**
	Examples of how the condition may be demonstrated — Evidence of what has been done, or is planned, to develop the experience of the senior regulators; — Proposals on the overall approach to assessment, licensing, inspection and enforcement, among other things; — Plans to further enhance the capability of the regulatory body for safety and security; — Plans to develop the required regulations, in line with IAEA safety standards and nuclear security recommendations; — Evidence of interaction and cooperation with established regulatory organizations; — Plans to secure assistance from international regulatory organizations or technical support organizations.
	Selected relevant IAEA publications Governmental, Legal and Regulatory Framework for Safety, IAEA General Safety Requirements No. GSR Part 1 (Rev.1), IAEA, Vienna (2016) Establishing the Infrastructure for Radiation Safety, Safety Standards Series No. SSG-44, IAEA, Vienna (2018) Objective and Essential Elements of a State's Nuclear Security Regime, NSS No. 20, IAEA, Vienna (2013) Nuclear Security Recommendations on Radioactive Material and Associated Facilities, NSS No. 14, IAEA, Vienna (2011) Security of Radioactive Material in Use and Storage and Associated Facilities, NSS No. 11-G (Rev.1), IAEA, Vienna (2019) Security of Radioactive Material in Transport, NSS No. 9-G (Rev. 1), IAEA, Vienna (2020)

8. Radiation protection [b]	
Conditions	**Basis for evaluation**
8.1. Existing radiation protection measures assessed and, if needed, enhancements to radiation protection programmes planned	**Summary of the condition to be demonstrated** The needed enhancements to the existing radiation protection programme to address IRF operation have been identified, including consideration of transport of radioactive materials and radioactive waste management. They consider both the increase in scale and the need to cover new technical issues. This issue is closely linked to infrastructure issue No. 7, regulatory framework.
	Examples of how the condition may be demonstrated — Evidence of discussions with specialists from other organizations or countries; — Identification of the main areas requiring enhancement; — Recognition that additional competencies will be required to review proposed designs against the requirement to control contamination and to optimize safety and protection in accordance with the radiation protection principles; — Recognition that the programme for dose assessment might need to be expanded; — Plans for who will be responsible for the radiation protection programme; — Provisions for competence and training of IRF staff on radiation safety and radiation protection.
	Selected relevant IAEA publications Radiation Protection of the Public and the Environment, Safety Standards Series No. GSG-8, IAEA, Vienna (2018) Radiation Protection and Safety of Radiation Sources: International Basic Safety Standards, Safety Standards Series No. GSR Part 3, IAEA, Vienna (2014) Radiation Protection and Safety in Medical Uses of Ionizing Radiation, Safety Standards Series No. SSG-46, IAEA, Vienna (2018) Radiation Safety of X Ray Generators and Other Radiation Sources Used for Inspection Purposes and for Non-medical Human Imaging, Safety Standards Series No. SSG-55, IAEA, Vienna (2020) Radiation Safety of Gamma, Electron and X Ray Irradiation Facilities, Safety Standards Series No. SSG-8, IAEA, Vienna (2010)

---

[b] This covers protection of workers and the public onsite during planned operation. Off-site releases from planned operation are addressed in infrastructure issue No. 13, environmental protection; accidental releases and associated radiation protection are addressed in infrastructure issue No. 14, emergency preparedness and response.

9. Utilization [c]	
Conditions	**Basis for evaluation**
9.1. Potential users and customers identified and consulted	**Summary of the condition to be demonstrated** The users and customers of an IRF are identified and consulted, their needs analysed and quantified. The mechanisms to adapt the IRF mission to evolving user and customer needs has been addressed.
	Examples of how the condition may be demonstrated — A justifiable list of the selected entities regarded as stakeholders, customers and users is available and is complete; — A methodology (e.g. questionnaire distributed and evaluated) for interaction between the future IRF operating organization and the identified stakeholders and users is developed; — A document reporting on the assessment of stakeholders' and users' needs has been prepared for initial evaluation regarding sufficiency and the proposed time schedule.
	Selected relevant IAEA publications IAEA Nuclear Energy Series No. NG-T-3.18, Feasibility Study Preparation for New Research Reactor Programmes, IAEA, Vienna (2018) IAEA Nuclear Energy Series No. NG-T-3.16, Strategic Planning for Research Reactors, IAEA, Vienna (2017)
9.2. Range of potential utilization of the IRF studied	**Summary of the condition to be demonstrated** The functional specifications for the IRF and associated infrastructure were developed on the basis of the assessment of stakeholder, user and customer needs. A full evaluation of the proposed initial capabilities of the IRF as well as the potential evolution during its lifetime are developed and documented.
	Examples of how the condition may be demonstrated — Documentation of the assessed stakeholder and user needs for the initial application of the IRF and associated infrastructure is available; — A further evaluation of the potential capabilities of the IRF as well as flexibility in design requirements to meet this are documented.

---

[c] This issue is closely linked to infrastructure issue No. 1, organizational position, condition No. 1.3. comprehensive feasibility study performed, documented and the necessary commitments understood, and issue No. 11, stakeholder involvement.

9. Utilization (cont.)	
Conditions	**Basis for evaluation**
	Selected relevant IAEA publications Feasibility Study Preparation for New Research Reactor Programmes, Nuclear Energy Series No. NG-T-3.18, IAEA, Vienna (2018) Strategic Planning for Research Reactors, Nuclear Energy Series No. NG-T-3.16, IAEA, Vienna (2017) Cyclotron Produced Radionuclides: Physical Characteristics and Production Methods, Technical Reports Series No. 468, IAEA, Vienna (2009) Neutron Generators for Analytical Purposes, Radiation Technology Reports No. 1, IAEA, Vienna (2012) Directory of Gamma Processing Facilities in Member States, IAEA, Vienna (2004) Planning National Radiotherapy Services: A Practical Tool, Human Health Series No. 14, IAEA, Vienna (2011)
9.3. The preliminary strategic planning document prepared	**Summary of the condition to be demonstrated** The rationale on the purpose and feasibility of the proposed IRF and its required auxiliaries and ancillaries have been presented, debated on, prioritized and accepted by the PMT.
	Example of how the condition may be demonstrated A draft PSP has been prepared and presented as part of the feasibility study for the proposed IRF and, among other items, includes: — Evaluation of stakeholder and user needs (immediate and future); — List of the identified and prioritized products and services of the IRF; — Identification of functional specification of the IRF and its associated infrastructure; — Role of the IRF in the regional and international contexts and the considerations for regional and international cooperation.
	Selected relevant IAEA publications Feasibility Study Preparation for New Research Reactor Programmes, Nuclear Energy Series No. NG-T-3.18, IAEA, Vienna (2018) Strategic Planning for Research Reactors, Nuclear Energy Series No. NG-T-3.16, IAEA, Vienna (2017) Establishing the Infrastructure for Radiation Safety, Safety Standards Series No. SSG-44, IAEA, Vienna (2018)

10. Human resource development	
Conditions	**Basis for evaluation**
10.1. Necessary knowledge and skills identified, and gaps in current capability assessed	**Summary of the condition to be demonstrated** A broad assessment of the typical staffing needs of each of the key organizations and their technical support has been completed together with an assessment of improvements required in the current capability of the country/organization to meet the project need. The assessment covers the full range of scientific, technical, managerial and administrative disciplines (both for occupancy and number of positions) and considers: — Current human resource competencies and capabilities; — Estimated required competence and capability, including qualification and certification, if applicable; — Estimated risk factors in knowledge preservation and transfer (e.g. personnel attrition, diminishing job tenures, decreasing availability of skills) and measures to address them; — Availability of domestic and foreign capacity for education and training; — Which facilities and programmes need to be established for education, training and experience building; — Which research, development and business capabilities need to be developed, if applicable; — A talent management programme, including for senior leader development.
	Examples of how the condition may be demonstrated — An analysis identifying the competencies and number of staff needed, covering all the relevant organizations and authorities. The analysis needs to include a breakdown by knowledge, skills and discipline. — An assessment of the capability of existing education and training facilities. — An assessment of knowledge loss risk (for operating facilities).
	Selected relevant IAEA publications Leadership and Management for Safety, Safety Standards Series No. GSR Part 2, IAEA, Vienna (2016) The Operating Organization and the Recruitment, Training and Qualification of Personnel for Research Reactors, Safety Standard Series No. NS-G-4.5, IAEA, Vienna (2008) Managing Human Resources in the Field of Nuclear Energy, Nuclear Energy Series Guides No. NG-G-2.1, IAEA, Vienna (2009) Knowledge Loss Risk Management in Nuclear Organizations, Nuclear Energy Series No. NG-T-6.11, IAEA, Vienna (2017) Establishing the Infrastructure for Radiation Safety, Safety Standards Series No. SSG-44, IAEA, Vienna (2018) Objective and Essential Elements of a State's Nuclear Security Regime, NSS No. 20, IAEA, Vienna (2013) Nuclear Security Recommendations on Radioactive Material and Associated Facilities, NSS No. 14, IAEA, Vienna (2011)

10. Human resource development (cont.)	
Conditions	**Basis for evaluation**
10.2. Development of human resources planned	**Summary of the condition to be demonstrated** Outline plans have been agreed to — Enhance national education and training, if needed; — Develop a detailed human resource development plan for each key organization and authority.

	Examples of how the condition may be demonstrated
	— Plans to develop human resources required including:
	   (a) Identification of national organizations that could support human resource development;
	   (b) Enhancement of education and training infrastructure;
	   (c) Development of national competencies (through schools, universities, institutes and industry);
	   (d) Non-national human resources that are needed to augment national resources and how they will be secured;
	   (e) International cooperation and vendor support;
	   (f) Leadership development;
	   (g) Succession planning.
	— Strategies for the recruitment and retention of staff;
	— Recognition of the need for qualification and certification programmes for personnel;
	— Evidence that key stakeholder organizations have participated in the development and review of the plans.

	Selected relevant IAEA publications Leadership and Management for Safety, Safety Standards Series No. GSR Part 2, IAEA, Vienna (2016) The Operating Organization and the Recruitment, Training and Qualification of Personnel for Research Reactors, Safety Standard Series No. NS-G-4.5, IAEA, Vienna (2008) Establishing the Infrastructure for Radiation Safety, Safety Standards Series No. SSG-44, IAEA, Vienna (2018) Managing Human Resources in the Field of Nuclear Energy, Nuclear Energy Series Guides No. NG-G-2.1, IAEA, Vienna (2009) Objective and Essential Elements of a State's Nuclear Security Regime, NSS No. 20, IAEA, Vienna (2013) Nuclear Security Recommendations on Radioactive Material and Associated Facilities, NSS No. 14, IAEA, Vienna (2011)

11. Stakeholder involvement [d]	
Conditions	**Basis for evaluation**
11.1. Stakeholder involvement programme initiated	**Summary of the condition to be demonstrated** The stakeholders of the IRF were identified and consulted. Stakeholder involvement strategy and plan, with the required resources and competence, initiated by the PIT with endorsement of the PMT. Information about the benefits and any potential risks of the IRF were communicated to all relevant and interested parties, including the general public.
	Examples of how the condition may be demonstrated — A clear mandate for the PIT to engage with stakeholders; — Actions to disseminate information in the context of the national nuclear science and technology road map, needs for IRF products and services, pros and cons for alternative technologies, using a range of effective tools; — Evidence of a professional communication team available to the PIT, with appropriate financial resources; — Approaches to address public concerns, including radioactive waste management and possible accidents; — Evidence of activities at the local, regional and national level; — A plan for ongoing interaction with the public, in particular, opinion leaders, media, local and national governmental officials.
	Selected relevant IAEA publications Feasibility Study Preparation for New Research Reactor Programmes, Nuclear Energy Series No. NG-T-3.18, IAEA, Vienna (2018) Leadership and Management for Safety, Safety Standards Series No. GSR Part 2, IAEA, Vienna (2016) Stakeholder involvement in nuclear Issues, INSAG-20, IAEA, Vienna (2006) Objective and Essential Elements of a State's Nuclear Security Regime, NSS No. 20, IAEA, Vienna (2013) Nuclear Security Recommendations on Radioactive Material and Associated Facilities, NSS No. 14, IAEA, Vienna (2011)

[d] This issue is closely linked to infrastructure issue No. 9, utilization, condition No. 9.1. users and customers identified and consulted.

12. Site survey, selection and evaluation	
Conditions	**Basis for evaluation**
12.1. If needed, survey of potential sites conducted, and candidate sites identified	**Summary of the condition to be demonstrated** Potential sites based on relevant selection criteria were considered, covering safety, security, cost, socioeconomic issues, engineering and the environment, and have been identified. Regional analysis to identify candidate sites has been conducted, if applicable. The analysis includes the impact of external hazards on security and emergency response capability. Consultations with stakeholders have been part of the process.

	Examples of how the condition may be demonstrated
	— A report covering:
	  (a) Safety and security criteria for initial IRF site selection;
	  (b) National criteria (e.g. socioeconomic and environmental);
	  (c) Engineering and cost criteria.
	— An assessment report issued and approved identifying:
	  (a) Regional analysis and identification of potential sites;
	  (b) Screening of potential sites and selection of candidate sites.
	— Evidence that the resources that were used for IRF site selection are competent and have experience with IRF site selection;
	— Plans for the work that will be required in Phase 2 to select and justify the site;
	— Evidence that safety and security related activities conducted (e.g. site evaluation and environmental impact studies) are included within the framework of an effective management system.

	Selected relevant IAEA publications Site Evaluation for Nuclear Installations, No. NS-R-3 (Rev. 1), IAEA, Vienna (2019) Site Survey and Site Selection for Nuclear Installations, No. SSG-35, IAEA, Vienna (2015)

13. Environmental protection [e]	
Conditions	**Basis for evaluation**
13.1. Environmental protection framework and related requirements considered	**Summary of the condition to be demonstrated** The PIT has reviewed the suitability of the State's existing framework for environmental protection. The PIT and PMT have considered the main environmental requirements related to the siting of the IRF, including land use, water use, air and water quality and the impacts of possible radioactive effluents, if applicable.
	Examples of how the condition may be demonstrated — Identification of the key requirements for siting, construction, commissioning, operation, decommissioning and site liberation; — Evidence of discussions by specialists with countries/organizations operating similar IRFs; — Evidence that the non-radiological environmental issues such as water use, transport of materials, disposal of hazardous waste, additional environmental monitoring requirements and the impact during all stages of a facility or activity have been considered and taken into account by the PMT; — Procedures developed for the elaboration, reporting and assessment of environmental studies for IRFs and associated infrastructure; — Public consultation of stakeholders; — Integration and coordination between conventional and radiological requirements regarding the environment.

---

[e] This covers off-site releases from planned operation and all other environmental issues. Protection of workers and the public on-site during planned operation are addressed in infrastructure issue No. 8, radiation protection. If relevant, accidental releases and radiation are addressed in infrastructure issue No. 14, EPR. In the aspects that concern the national framework, significant gaps are not expected to exist, and this issue is usually checked by referring to the results of IAEA review missions and advisory services or by referring to existing IAEA databases and other electronic information resources. A list of these services and resources is given in Annex II.

14. Emergency preparedness and response	
Conditions	**Basis for evaluation**
14.1. Requirements of, and resources for, developing an emergency response capability recognized	**Summary of the condition to be demonstrated** The PMT is aware of the EPR arrangements and capabilities that might be required for the IRF project. It has evaluated existing EPR arrangements and capabilities in the organization developing the IRF project and is aware of the major gaps that will need to be addressed. The PMT has identified the main organizations and resources that will need to be involved in the establishment of adequate EPR capabilities. The lead for the execution of the action plan and the action plan coordination framework has been identified. If needed, the process of developing adequate EPR will be initiated in Phase 2 and will be largely carried out in Phase 3.
	Example of how the condition may be demonstrated Report summarizing existing EPR arrangements and capabilities and identifying those to be enhanced and/or developed as well as identifying the main organizations and resources that will need to be involved in the establishment of adequate national EPR capabilities.
	Selected relevant IAEA publications Preparedness and Response for a Nuclear or Radiological Emergency, Safety Standard Series No. GSR Part 7, IAEA, Vienna (2015) Criteria for Use in Preparedness and Response for a Nuclear or Radiological Emergency, Safety Standard No. GSG-2, IAEA, Vienna (2011) Arrangements for Preparedness for a Nuclear or Radiological Emergency, Safety Standard No. GS-G-2.1, IAEA, Vienna (2007) Objective and Essential Elements of a State's Nuclear Security Regime, NSS No. 20, IAEA, Vienna (2013) Nuclear Security Recommendations on Radioactive Material and Associated Facilities, NSS No. 14, IAEA, Vienna (2011)

15. Nuclear security [a]	
Conditions	**Basis for evaluation**
15.1. Security requirements recognized, and the actions of all relevant organizations coordinated	**Summary of the condition to be demonstrated** The PMT and PIT recognize the importance of the security of nuclear and/or radioactive materials owing to an organizational threat assessment for the organization/institution/company promoting the IRF and principles of prevention, detection and response. All competent authorities that are involved in nuclear and/or radioactive material security have been identified, and there is a coordinating body or mechanism established that brings together all the organizations that have responsibility for security of such materials. The need to establish legislation and a regulatory framework is addressed under infrastructure issues No. 5 and 7, legal framework and regulatory framework, respectively.
	Examples of how the condition may be demonstrated — Evidence of familiarity with IAEA nuclear security series publications and other States' practices; — Clear identification of all organizations that have roles and responsibilities for security and of the work that will need to be carried out in the subsequent phases; — Evidence that nuclear security considerations for siting have been defined and have been considered as part of the siting assessment (see infrastructure issue No. 12, site survey, selection and evaluation); — Evidence that international cooperation and assistance is being used; — Evidence that the need to address the interface with safety is recognized.
	Selected relevant IAEA publications Objective and Essential Elements of a State's Nuclear Security Regime, NSS No. 20, IAEA, Vienna (2013) Nuclear Security Recommendations on Radioactive Material and Associated Facilities, NSS No. 14, IAEA, Vienna (2011) Security of Radioactive Material in Use and Storage and Associated Facilities, NSS No. 11-G (Rev. 1), IAEA, Vienna (2019) Security of Radioactive Material in Transport, NSS No. 9-G (Rev. 1), IAEA, Vienna (2020) Code of Conduct on the Safety and Security of Radioactive Sources, IAEA/CODEOC/2004, IAEA, Vienna (2004) Guidance on the Import and Export of Radioactive Sources, IAEA/CODEOC/IMO-EXP/2012, IAEA, Vienna (2012) Guidance on the Management of Disused Radioactive Sources,

---

[a] A condition to develop a successful IRF project is that no major gaps in the infrastructure development for this issue exist, as the national framework is, in most cases, not expected to require significant changes for the purpose of establishing an IRF. This issue is usually checked by referring to the results of IAEA review missions and advisory services or by referring to existing IAEA databases and other electronic information resources. A list of these services and resources is given in Annex II.

16. Fostering expansion of ownership	
Conditions	**Basis for evaluation**
16.1. Potential for fostering expansion of IRF ownership through technology transfer support considered, where applicable, and relevant/candidate recipient entities identified	**Summary of the condition to be demonstrated** Evidence of interest in adoption of IRF technology transfer support exists, and criteria for being an eligible recipient have been identified. The analysis to assess readiness of potential candidate entities, including private sector, has been conducted; it includes *inter alia* past professional history, current functions/operations and their management, compliance with quality management systems and applicable safety standards, adherence to safeguards, safety and security responsibilities, demonstrable commitment covering the foreseen life cycle of the IRF, and similar.

	Examples of how the condition may be demonstrated
	— A report covering:
	    (a) Survey conducted to assess evidence of interest (both extent and volume) in adopting IRF by technology transfer;
	    (b) National laws and procedures governing ownership and management of IRF operations/services (if existent).
	— An analysis and assessment report issued identifying:
	    (a) Eligibility criteria for entity(ies) being recipient(s);
	    (b) Fulfilment information template for potential recipient(s) developed and available;
	    (c) Screening methods of potential recipient(s) and selection of candidate recipient(s).
	— Evidence (to demonstrate) that the technology and competency resources required for IRF ownership are adequately understood and available with recipient(s) or can be acquired in a time bound manner;
	— Plans for the work that will be required in Phase 2 to advance further with the expansion of IRF ownership by technology transfer support.

	Selected relevant IAEA publications Leadership and Management for Safety, Safety Standards Series No. GSR Part 2, IAEA, Vienna (2016)

17. Radioactive waste management	
Conditions	**Basis for evaluation**
17.1. Requirements for management of radioactive waste from IRF recognized	**Summary of the condition to be demonstrated** The PMT and PIT understand the increased requirements for the processing, storage and disposal of radioactive waste from IRFs and have developed options for managing it safely, taking into account existing arrangements. The PMT and PIT understand the options for the management of each of the different waste categories, including the return of disused radioactive sources to the manufacturer or the supplier, in consideration of the capacity for radioactive waste management in the country. Although the specific routes for disposal of the different waste categories can be decided later, the need to select and plan for adequate options is recognized.
	Example of how the condition may be demonstrated A document addressing possible approaches to the management of radioactive waste arising from IRF operation and decommissioning, the capabilities and resources needed, and the options and technologies for its processing, handling, storage and disposal. Such a document should also include possible options for disposal of different radioactive waste categories and options for funding these activities. Insurance or pre-pay of disposal is arranged. Regulatory framework and financing schemes are addressed under infrastructure issues No. 7, regulatory framework, and No. 4 funding and financing.
	Selected relevant IAEA publications IAEA Policies and Strategies for Radioactive Waste Management, No. NW-G-1.1, IAEA, Vienna (2009) Policies and Strategies for the Decommissioning of Nuclear and Radiological Facilities, No. NW-G-2.1, IAEA, Vienna (2011) Predisposal Management of Radioactive Waste, Safety Standards Series No. GSR Part 5, IAEA, Vienna (2009) Decommissioning of Facilities, Safety Standards Series No. GSR Part 6, IAEA, Vienna (2014) Decommissioning of Small Medical, Industrial and Research Facilities: A Simplified Stepwise Approach, Nuclear Energy Series No. NW-T-2.3, IAEA, Vienna (2011) IAEA Management of Disused Sealed Radioactive Sources, Nuclear Energy Series No. NW-T-1.3, IAEA, Vienna (2014) IAEA Decommissioning of Particle Accelerators, Nuclear Energy Series No. NW-T-2.9, IAEA, Vienna (2020) Nuclear Security Recommendations on Radioactive Material and Associated Facilities, NSS No. 14, IAEA, Vienna (2011) Code of Conduct on the Safety and Security of Radioactive Sources, IAEA/CODEOC/2004, IAEA, Vienna (2004) Guidance on the Import and Export of Radioactive Sources, IAEA/CODEOC/IMO-EXP/2012, IAEA, Vienna (2012) Guidance on the Management of Disused Radioactive Sources, IAEA/CODEOC/MGT-DRS/2018, IAEA, Vienna (2018)

18. Industrial involvement	
Conditions	**Basis for evaluation**
18.1. Involvement of national/domestic industrial and utility services considered	**Summary of the condition to be demonstrated** A recommended policy for national involvement, covering availability of expertise, industrial capability and technical/utility services for the IRF being considered. The balance between capability, quality standards and intended industrial development is recognized, and if required, is in compliance with national policy.
	Examples of how the condition may be demonstrated — A survey of companies with the potential to participate in the IRF project for construction, equipment provision or support services, with a review of their ability to satisfy the requirements of the IRF project; — Meetings with, or training of, potential suppliers to explain standards and qualifications required, review feasibility of involvement and identify required actions and funding requirements.
	Selected relevant IAEA publications Feasibility Study Preparation for New Research Reactor Programmes, Nuclear Energy Series No. NG-T-3.18, IAEA, Vienna (2018) Technical Requirements in the Bidding Process for a New Research Reactor, Nuclear Energy Series No. NP-T-5.6, IAEA, Vienna (2014) Leadership and Management for Safety, Safety Standards Series No. GSR Part 2, IAEA, Vienna (2016)

19. Procurement	
Conditions	**Basis for evaluation**
19.1. Requirements and procedures for purchasing IRF recognized	**Summary of the condition to be demonstrated** Recognition by the organization/institution/company of the requirements and procedures associated with purchasing IRFs and associated facilities.
	Examples of how the condition may be demonstrated — Appropriate procurement of consulting services in Phase 1, if needed; — Evidence that the issues related to services for Phase 2 activities are recognized, allowing for both national and foreign suppliers.
	Selected relevant IAEA publications Feasibility Study Preparation for New Research Reactor Programmes, Nuclear Energy Series No. NG-T-3.18, IAEA, Vienna (2018) Technical Requirements in the Bidding Process for a New Research Reactor, Nuclear Energy Series No. NP-T-5.6, IAEA, Vienna (2014) Leadership and Management for Safety, Safety Standards Series No. GSR Part 2, IAEA, Vienna (2016)

Annex II

IAEA RESOURCES

This annex lists IAEA review missions and advisory services as well as other databases, electronic information resources and tools that are relevant to the assessment of the infrastructure issues for a new IRF project.

The types of IRF targeted by this guide are numerous as well as diverse (see Section 1.4 for details). Therefore, it is recommended that the list of services in this annex be considered by applying due care and keeping in mind the specific nature of the IRF under consideration, and also applying a graded approach and tailoring it as appropriate to the nature and complexity of the IRF.

II–1. IAEA LEGISLATIVE ASSISTANCE PROGRAMME

The IAEA Legislative Assistance Programme [II–1] covers all branches of international and national nuclear law: nuclear safety; nuclear security; safeguards and non-proliferation; and liability for nuclear damage. Legislative assistance is available on request to all Member States, regardless of the extent of their nuclear activities. The programme is implemented through: bilateral legislative drafting assistance and reviews of enacted legislation to assist States in complying with their international obligations and commitments, as well as in implementing relevant IAEA safety standards and nuclear security guidance; awareness raising missions; national and regional workshops and training courses and seminars; and the training of individuals including at the annual Nuclear Law Institute.

II–2. IAEA REVIEW MISSIONS AND ADVISORY SERVICES

The IAEA offers its Member States a wide array of review services, in which an IAEA-led team of experts compares actual practices with IAEA standards and guidance IAEA [II–2]. These review missions and advisory services are available on request by the Member State, and each covers a specific area and has its own objectives. Several of them are strongly connected to one or more infrastructure issues, and as such can be used by Member States to assist and guide them in their self-assessment of infrastructure issues for a new IRF project.

II–2.1. Independent Safety Culture Assessment

The purpose of an Independent Safety Culture Assessment (ISCA) [II–3] is to support high levels of safety by agreeing upon a common view of an organization's safety culture and identifying areas where improvements can be made.

II–2.2. Safety Culture Continuous Improvement Process

The Safety Culture Continuous Improvement Process (SCCIP) [II–4] assists Member States in strengthening and maintaining their safety culture. The process involves training that enables staff from a receiving organization to use safety culture assessments to improve safety culture, to develop effective improvement programmes and to create effective and sustainable organizational change. The trained staff members, who should represent different departments and different levels, form a safety culture improvement team with in-house expertise.

II–2.3. Transport Safety Appraisal Service

The Transport Safety Appraisal Service (TranSAS) [II–5] reviews the implementation of transport regulations to enable Member States to achieve a high level of safety in the transport of radioactive material. The service provides recommendations for improvement where appropriate.

II–2.4. Integrated Regulatory Review Service

The Integrated Regulatory Review Service (IRRS) [II–6] offers an integrated approach to the review of common aspects of any State's national, legal and governmental framework and regulatory infrastructure for safety. The IRRS provides a peer review of both regulatory technical and policy issues and is suitable for any State, regardless of the level of development of its activities and practices that involve ionizing radiation or a nuclear programme. IRRS teams evaluate a State's regulatory infrastructure for safety against IAEA safety standards.

II–2.5. Advisory Mission on Regulatory Infrastructure for Radiation Safety and Nuclear Security

The Advisory Mission on Regulatory Infrastructure for Radiation Safety and Nuclear Security (known as an RISS) is conducted in countries where significant actions are necessary for the country's regulatory infrastructure to meet the provisions of the IAEA safety standards and nuclear security guidance, the Code of Conduct on the Safety and Security of Radioactive Sources and its Supplementary Guidance [20]. An RISS might be conducted in States with essentially no regulatory infrastructure for radiation safety or nuclear security. Advisory missions conducted in accordance with these guidelines address the national regulatory infrastructure for radiation safety and nuclear security and, therefore, the regulatory infrastructure for nuclear and fuel cycle facilities is outside their scope. Regulatory requirements for radiological EPR (as indicated in the Code of Conduct for the Safety and Security of Radioactive Sources) may be included in the scope of the mission in addition, upon request by the host country.

II–2.6. Occupational Radiation Protection Appraisal Service

Occupational Radiation Protection Appraisal Service (known as ORPAS) missions [II–7] are conducted as an independent appraisal service in the field of radiation protection of workers. Such an appraisal is an opportunity for a Member State to have all or some aspects of its occupational radiation protection programme independently assessed and evaluated against international safety standards.

II–2.7. Emergency Preparedness Review

The Emergency Preparedness Review (EPREV) service [II–8] is provided by the IAEA to Member States on their request to appraise their level of preparedness for nuclear or radiological emergencies.

II–2.8. Integrated Review Service for Radioactive Waste and Spent Fuel Management, Decommissioning and Remediation

The Integrated Review Service for Radioactive Waste and Spent Fuel Management, Decommissioning and Remediation (ARTEMIS) [II–9] is an integrated expert peer review service for radioactive waste and spent fuel management, decommissioning and remediation programmes. This service is intended for facility operators and organizations responsible for radioactive waste management, as well as for regulators, national policy and other decision makers.

II–2.9. International Physical Protection Advisory Service

The International Physical Protection Advisory Service (IPPAS) [II–10], created by the IAEA in 1995, provides peer advice on implementing international instruments and Agency guidance on the protection of nuclear and other radioactive material, associated facilities and associated activities.

An IPPAS mission compares a State's existing practices against relevant international instruments and IAEA nuclear security publications. It also includes an exchange of experience and good international practices aimed at strengthening the State's nuclear security regime. IPPAS missions comprise a national level review of the legal and regulatory framework. Depending on a State's request, they may also include a review of security systems and measures at facilities and during the transport of nuclear and other radioactive material. IPPAS missions also can cover computer security.

IPPAS missions are conducted by teams of international nuclear security experts. Team members use their extensive experience and international guidance to suggest improvements. Conclusions are made by consensus on the basis of the team's combined expertise.

The findings of IPPAS missions are reflected in mission reports, which are treated by the IAEA as highly confidential. Upon request, the missions can be complemented by IAEA follow-up assistance, including training, technical support and more targeted assessments of various elements of a State's national nuclear security regime.

II–3. IAEA INFORMATION RESOURCES AND OTHER TOOLS

The IAEA provides access to a number of resources that are relevant to the assessment of the infrastructure issues for a new IRF project through its NUCLEUS portal [II–11], including databases, applications, publications and training material. Other tools available to Member States are also listed here.

II–3.1. Emergency preparedness and response resources

Through its Incident and Emergency Centre (known as IEC), the IAEA maintains four online tools and mechanisms [II–12] to share information and data on EPR for nuclear and radiological incidents and emergencies.

II–3.2. Global Nuclear Safety and Security Network

The IAEA Global Nuclear Safety and Security Network (GNSSN) [II–13] is both a human network and a web based platform, allowing its members to share nuclear safety and security knowledge and services to further the goal of achieving worldwide implementation of a high level of nuclear safety and security.

II–3.3. Information resources in radiation protection

This resource includes five sites that provide access to information on occupational exposure in medicine, industry and research; safety in radiation oncology and in radiological procedures; radiation protection of patients; and to occupational radiation protection networks [II–14].

II–3.4. Radiation Safety Information Management System

IAEA's Radiation Safety Information Management System (RASIMS) [II–15] enables Member States to assess how closely their infrastructures for radiation safety are aligned with IAEA safety standards requirements.

RASIMS is a web based platform that gives Member States the framework for collecting, analysing and viewing information that reflects the status of their national infrastructure for radiation safety. The system is specifically aimed at supporting those IAEA Member States that receive technical assistance from the Agency.

The information in RASIMS is grouped into thematic safety areas (TSAs) that are derived from the IAEA's safety standards.

II–3.5. Self-Assessment of Regulatory Infrastructure for Safety

Self-assessment aims to improve the effectiveness and efficiency of the regulatory body and its activities. Through the module for regulatory bodies of eSARIS of the resource Self-Assessment of Regulatory Infrastructure for Safety (SARIS) [II–16], regulatory body staff can reflect on important aspects of regulatory policy and strategy to achieve the fundamental safety objective by enhancing regulatory processes and performance. Self-assessment enables users to verify whether a regulatory body performs its functions in line with the IAEA safety standards and whether an organizational culture for continuous improvement is sufficiently maintained.

II–3.6. Regulatory Authority Information System

The Regulatory Authority Information System (RAIS) [II–17] is a software application developed by the IAEA to assist the regulatory bodies of Member States in managing their regulatory control programmes in accordance with IAEA safety standards and guides and in accordance with the Code of Conduct on the Safety and Security of Radioactive Sources and its supplementary guidance.

The main features of the Regulatory Authority Information System are the maintenance of registries and records of regulatory data, the management of regulatory information and the management of regulatory activities.

II–3.7. Systematic Assessment of Regulatory Competence Needs

The IAEA-developed Guidelines for Systematic Assessment of Regulatory Competence Needs (SARCoN) [II–18] methodology and software tool support Member States' work to ensure regulatory competence in line with the IAEA safety standards.

The SARCoN methodology and tool support the competence management process in regulatory bodies for nuclear facilities and/or for radiation sources facilities and activities. It assists in evaluation and self-assessment processes through a step-by-step approach that enables users to develop competence profiles and conduct competence gap analyses. Regulatory bodies can use the methodology and tool to ensure that their current and future competence needs are met.

II–3.8. Nuclear Security Information Management System

The Nuclear Security Information Management System (known as NUSIMS) is a voluntary system designed to assist Member States in reviewing the status of their nuclear security infrastructure through self-assessment, as well as in tracking their progress towards establishing, maintaining and sustaining an effective nuclear security regime. The NUSIMS facilitates the systematic identification and prioritization of their nuclear security needs and allows the IAEA, **upon request,** to provide a more tailored approach in addressing those needs, all in one system.

The structure of the system, including the nuclear security areas and themes, was developed and validated using the following:

— IAEA nuclear security series publications;
— Input from Member State experts.

II–3.9. Integrated Nuclear Security Support Plan

The Integrated Nuclear Security Support Plan (INSSP) [II–19] is a non-legally binding document; however, its approval or endorsement by the State is considered to be indicative of a commitment to pursue the implementation of agreed nuclear security improvements contained therein. States that receive IAEA assistance in nuclear security through the INSSP process are not required to contribute to the Nuclear Security Fund.

The overall objectives of the INSSP are to identify and consolidate the nuclear security needs of an individual State into an integrated document that includes the necessary nuclear security improvements, as well as to provide a customized framework for coordinating and implementing nuclear security activities conducted by the State, the IAEA and potential donors. The INSSP is designed to identify actions required to ensure that a State's national nuclear security regime is effective and sustainable, based on IAEA nuclear security guidance.

States that are interested in cooperating with the IAEA to develop an INSSP can address an **official request** to the Director of the IAEA Nuclear Security Division. This can be preceded, if necessary and upon request, by a national visit from experts of the IAEA to sensitize senior level officials to the importance of nuclear security and the INSSP process.

REFERENCES TO ANNEX II

[II–1] Legislative Assistance,
        https://www.iaea.org/services/legislative-assistance
[II–2] Review Missions and Advisory Services,
        https://www.iaea.org/services/review-missions
[II–3] Independent Safety Culture Assessment,
        https://www.iaea.org/services/review-missions/independent-safety-culture-assessment-isca
[II–4] Safety Culture Continuous Improvement Process,
        https://www.iaea.org/services/review-missions/safety-culture-continuous-improvement-process-sccip
[II–5] Transport Safety Appraisal Services,
        https://www.iaea.org/services/review-missions/transas
[II–6] Integrated Regulatory Review Service (IRRS),
        https://www.iaea.org/services/review-missions/integrated-regulatory-review-service-irrs
[II–7] Occupational Radiation Protection Appraisal Service (ORPAS),
        https://www.iaea.org/services/review-missions/occupational-radiation-protection-appraisal-service-orpas
[II–8] Emergency Preparedness Review (EPREV) Service,
        https://www.iaea.org/services/review-missions/emergency-preparedness-review-eprev-service
[II–9] Integrated Review Service for Radioactive Waste and Spent Fuel Management, Decommissioning and Remediation (ARTEMIS),
        https://www.iaea.org/services/review-missions/integrated-review-service-for-radioactive-waste-and-spent-fuel-management-decommissioning-and-remediation-artemis

[II–10]   International Physical Protection Advisory Service (IPPAS),
          https://www.iaea.org/services/review-missions/international-physical-protection-advisory-service-ippas

[II–11]   NUCLEUS Information Resources,
          https://www.iaea.org/resources/nucleus-information-resources

[II–12]   Incident and Emergency Centre (IEC), Emergency Preparedness and Response Resources,
          https://www.iaea.org/resources/databases/emergency-preparedness-and-response-resources

[II–13]   Global Nuclear Safety and Security Network (GNSSN),
          https://www.iaea.org/services/networks/global-nuclear-safety-and-security-network

[II–14]   Information Resources in Radiation Protection,
          https://www.iaea.org/resources/databases/radiation-protection

[II–15]   Radiation Safety Information Management System (RASIMS),
          https://www.iaea.org/resources/databases/rasims

[II–16]   Self-Assessment of Regulatory Infrastructure for Safety (SARIS),
          https://www.iaea.org/resources/software/e-saris

[II–17]   Regulatory Authority Information System (RAIS),
          https://www.iaea.org/resources/software/rais

[II–18]   Systematic Assessment of Regulatory Competence Needs (SARCoN),
          https://www.iaea.org/resources/software/systematic-assessment-of-regulatory-competence-needs-sarcon

[II–19]   Integrated Nuclear Security Support Plan (INSSP), IAEA, Vienna,
          https://www.iaea.org/topics/integrated-nuclear-security-support-plan-inssp

ABBREVIATIONS

EPR	emergency preparedness and response
FSR	feasibility study report
IAR	infrastructure assessment report
INSSP	Integrated Nuclear Security Support Plan
IPPAS	International Physical Protection Advisory Service
IRF	ionizing radiation facility
IRRS	Integrated Regulatory Review Service
NNI	national nuclear institution
PIT	project implementation team
PMT	project management team
PSP	preliminary strategic plan
R&D	research and development
SER	self-evaluation report

CONTRIBUTORS TO DRAFTING AND REVIEW

Alexeeva, V.	International Atomic Energy Agency
Andrada Contardi, F.	International Atomic Energy Agency
Antariksawan, A.	National Nuclear Energy Agency, Indonesia
Apperley, M.	Australian Nuclear Science and Technology Organisation, Australia
Bastos, J.	International Atomic Energy Agency
Bilic Zabric, T.	International Atomic Energy Agency
Blackburn, C.M.	International Atomic Energy Agency
Bode, P.	Consultant, Netherlands
Charisopoulos, S.	International Atomic Energy Agency
Cherf, A.	Consultant, Algeria
Cruz Suarez, R.	International Atomic Energy Agency
Doo, J.Y.	International Atomic Energy Agency
Ek, D.	International Atomic Energy Agency
El-Haj, Noura	International Atomic Energy Agency
Gaspar, M.	International Atomic Energy Agency
Han, B. S.	International Atomic Energy Agency
Hayes, T.	International Atomic Energy Agency
Horvath, K.	International Atomic Energy Agency
Kamenopoulou, V.	International Atomic Energy Agency
Khaliq, M.	International Atomic Energy Agency
Kovachev, M.	International Atomic Energy Agency
Lim, I.-C.	Korea Atomic Energy Research Institute, Republic of Korea

Ljubenov, V.	International Atomic Energy Agency
Luis Hernandez, J.	International Atomic Energy Agency
Mavric, H.	International Atomic Energy Agency
Michal, V.	International Atomic Energy Agency
Mittendorfer, J.	Consultant, Austria
Monaghan, B.	Consultant, Canada
Osso, J.A., Jr.	International Atomic Energy Agency
Pessoa Barradas, N.	International Atomic Energy Agency
Ramamoorthy, N.	Consultant, India
Ranguelova, V.	International Atomic Energy Agency
Ridikas, D.	International Atomic Energy Agency
Rodriguez y Baena, A.	International Atomic Energy Agency
Sharma, R.	International Atomic Energy Agency
Simon, A.	International Atomic Energy Agency
Skukan, N.	International Atomic Energy Agency
Sunshine, A.	International Atomic Energy Agency
Swainson, I.	International Atomic Energy Agency
Syahril, S.	International Atomic Energy Agency
Tanboon, N.	Thailand Institute of Nuclear Technology, Thailand
Tonos Paniagua, F.C.	International Atomic Energy Agency
Tuấn, H.A.	Vietnam Atomic Energy Agency, Vietnam
Waseem, M.	International Atomic Energy Agency
Wetherall, A.	International Atomic Energy Agency

Consultancy Meeting
Vienna, Austria: 21–24 October 2019

Regional Workshop
Vienna, Austria: 21–22 October 2020

ORDERING LOCALLY

IAEA priced publications may be purchased from the sources listed below or from major local booksellers.

Orders for unpriced publications should be made directly to the IAEA. The contact details are given at the end of this list.

NORTH AMERICA

Bernan / Rowman & Littlefield
15250 NBN Way, Blue Ridge Summit, PA 17214, USA
Telephone: +1 800 462 6420 • Fax: +1 800 338 4550
Email: orders@rowman.com • Web site: www.rowman.com/bernan

REST OF WORLD

Please contact your preferred local supplier, or our lead distributor:

Eurospan Group
Gray's Inn House
127 Clerkenwell Road
London EC1R 5DB
United Kingdom

Trade orders and enquiries:
Telephone: +44 (0)176 760 4972 • Fax: +44 (0)176 760 1640
Email: eurospan@turpin-distribution.com

Individual orders:
www.eurospanbookstore.com/iaea

For further information:
Telephone: +44 (0)207 240 0856 • Fax: +44 (0)207 379 0609
Email: info@eurospangroup.com • Web site: www.eurospangroup.com

Orders for both priced and unpriced publications may be addressed directly to:
Marketing and Sales Unit
International Atomic Energy Agency
Vienna International Centre, PO Box 100, 1400 Vienna, Austria
Telephone: +43 1 2600 22529 or 22530 • Fax: +43 1 26007 22529
Email: sales.publications@iaea.org • Web site: www.iaea.org/publications